THE OXFORD ENGINEERING SCIENCE SERIES

General Editors

J. M. BRADY   A. L. CULLEN   T. V. JONES
J. VAN BLADEL   L. C. WOODS   C. P. WROTH

UNOTTSCI

D1356364

THE OXFORD ENGINEERING SCIENCE SERIES

Boiling, Condensation, and Gas–Liquid Flow

P. B. WHALLEY

Department of Engineering Science,
University of Oxford

CLARENDON PRESS · OXFORD

Oxford University Press, Walton Street, Oxford OX2 6DP
Oxford New York Toronto
Delhi Bombay Calcutta Madras Karachi
Petaling Jaya Singapore Hong Kong Tokyo
Nairobi Dar es Salaam Cape Town
Melbourne Auckland
and associated companies in
Berlin Ibadan

Oxford is a trade mark of Oxford University Press

Published in the United States
by Oxford University Press, New York

© P. B. Whalley, 1987

First published 1987
First published in paperback 1990

All rights reserved. No part of this publication may be reproduced,
stored in a retrieval system, or transmitted, in any form or by any means,
electronic, mechanical, photocopying, recording, or otherwise, without
the prior permission of Oxford University Press

This book is sold subject to the condition that it shall not, by way
of trade or otherwise, be lent, re-sold, hired out or otherwise circulated
without the publisher's prior consent in any form of binding or cover
other than that in which it is published and without a similar condition
including this condition being imposed on the subsequent purchaser

British Library Cataloguing in Publication Data
Whalley, P. B.
Boiling, condensation, and gas–liquid flow.
—(Oxford engineering science series; 21)
1. Ebullition 2. Heat—Transmission
I. Title
536'.25 QC304
ISBN 0-19-856181-4
ISBN 0–19–856234–9 (Pbk)

Library of Congress Cataloging in Publication Data
Whalley, P. B.
Boiling, condensation, and gas–liquid flow
(Oxford engineering science series; 21)
Bibliography: p.
Includes index.
1. Ebullition. 2. Condensation. 3. Two-phase flow.
I. Title. II. Series.
TP363.W45 1987 660.2'8425 86-23731
ISBN 0-19-856181-4
ISBN 0–19–856234–9 (Pbk)

Typeset and printed by The Universities Press (Belfast) Ltd

UNIVERSITY
LIBRARY
NOTTINGHAM

PREFACE

This book is intended for final year undergraduates or graduate students in Mechanical Engineering or Chemical Engineering and is relevant to both the power and the process industries. It has been designed as an introduction to the subjects of two-phase gas–liquid flow, boiling, and condensation. The aim has been to present fairly basic information together with recommendations about, and full details of, some actual calculation procedures. References to source material and to more detailed texts are given at the end of each chapter.

I hope this book will go a little way to counteract the current over-academic approach to heat transfer which is evidenced by recent textbooks. Many of these books concentrate on heat transfer by conduction and radiation. Convective heat transfer is covered but only in idealized single-phase flow cases; boiling and condensation seem to be treated almost as an afterthought and in a few pages. Many, perhaps the majority, of the real problems in industrial heat transfer are concerned with changes of phase: boiling or condensation. As a necessary introduction to convective boiling and condensation, it is first necessary to know a reasonable amount about gas–liquid two-phase flow. These considerations dictated the layout of this book. In addition, at the end of the book there are two chapters about some of the more practical aspects of boiling and condensation equipment.

Some readers may consider that annular two-phase flow is over-emphasized. I can only reply that annular flow phenomena are interesting, amenable to study, and show the detailed types of results which can be obtained by modelling a particular type of flow. I need hardly add that annular flow has been my main research interest for a number of years.

I should particularly thank: Janet Vaughan and Barbara Granito for typing the original version of this book; the graduate students of the Department of Chemical Engineering of the University of Waterloo whose comments have been very useful; and Dr G. F. Hewitt of Harwell, from whom I have learnt so much.

Oxford   P.B.W.
1986

CONTENTS

1

INTRODUCTION

1.1 Two-phase flow and heat transfer

Two-phase flows are commonly found in industrial processes and in ordinary life. The phases may of course be solid, liquid, or gas.

Gas–liquid flow occurs in boiling and condensation operations, and inside many pipelines which nominally carry oil or gas alone, but which actually carry a mixture of oil and gas.

Liquid–liquid flow (i.e. flow of two immiscible liquids) occurs, for example, in liquid–liquid extraction processes.

Gas–solid flow occurs in a fluidized bed and in the pneumatic conveying of solid particles.

Solid–liquid flow occurs during the flow of suspensions such as river bed sediments and coal–water slurry.

This book is concerned exclusively with gas–liquid flow, although some of the principles and methods can be applied to other types of two-phase flow. The first part of the book deals with adiabatic flow, that is flow with no heat addition or removal. This generally means that the gas and liquid flow rates are constant, although in high speed flow (as in critical flow) partial vaporization of the liquid may occur even though there is no heat addition. The remainder of the book deals with heat transfer in two-phase situations: boiling and condensation. The objective here is to identify the various types of heat transfer and to show the heat transfer rate can be calculated.

Although it is not the primary objective of this book to specify calculation methods, a number of methods are recommended. These are often too cumbersome to fit neatly into the main text, so they are given as appendices.

1.2 Units

Throughout the text the SI system of units is used. Therefore almost all the equations are dimensionally consistent and there is no need for the 'gravitational constant' g_c. On the rare occasions when dimensionally incorrect equations are used the units of the variables in the equation are clearly specified. Experimental results are also quoted in SI units, the only exception is that the 'bar' is often used as the unit of pressure when displaying results. Of course 1 bar $= 10^5 \, \text{N/m}^2$.

1.3 Nomenclature

The symbols used are defined when they are introduced. However, some of the more common symbols are:

those for physical properties:

ρ density (kg/m^3),
μ viscosity (Ns/m^2),
k thermal conductivity (W/m K),
C_p specific heat (at constant pressure) (J/kg K);

subscripts:

l liquid phase,
g gas (or vapour) phase;

and those for flow variables:

G total mass flux (kg/m^2 s) = total (liquid + gas) mass flow rate (kg/s) divided by total channel cross-sectional area (m^2);
G_l liquid mass flux (kg/m^2 s) = liquid mass flow rate (kg/s) divided by total channel cross-sectional area (m^2);
G_g gas mass flux (kg/m^2 s) = gas mass flow rate (kg/s) divided by total channel cross-sectional area (m^2);
x quality—this is the fraction of the mass flow rate which is gas, and so

$$x = \frac{G_g}{G} \tag{1.1}$$

(quality is also sometimes referred to as the mass dryness fraction);

V_g superficial velocity of the gas (m/s)—this is the velocity if the gas in the two-phase flow was flowing alone in single phase flow in the channel, so

$$V_g = \frac{G_g}{\rho_g}; \tag{1.2}$$

V_l superficial velocity of the liquid (m/s), similarly defined, so

$$V_l = \frac{G_l}{\rho_l}; \tag{1.3}$$

α void fraction—this is the time-averaged fraction of the cross-sectional area or of the volume which is occupied by the gas phase;
u_g actual velocity of the gas phase (m/s)—this is the velocity which would be measured if the velocity of a small volume of gas could actually

be determined, so

$$u_g = \frac{V_g}{\alpha};$$ (1.4)

u_l actual velocity of the liquid phase (m/s), so

$$u_l = \frac{V_l}{1 - \alpha}.$$ (1.5)

1.4 Scope of the text

The scope of this book is inevitably limited as it is intended to be an introductory text. For further information the reader is referred in particular to the following books.

(*a*) Hetsroni (1982) for a detailed survey of almost all aspects of two-phase flow and heat transfer.

(*b*) Collier (1981) for a good summary of the available information about convective boiling.

(*c*) Wallis (1969) for a wide coverage of topics in adiabatic two-phase flow.

(*d*) Hewitt and Hall Taylor (1971) for a detailed description of annular two-phase flow.

(*e*) Hewitt (1978) for a description of instrumentation and flow measurement techniques for two-phase flow.

Later chapters in this book refer to specific parts of these books, other books, research papers, and reports. Each chapter is concluded with a list of references for that chapter.

References

COLLIER, J. G. (1981). *Convective boiling and condensation.* McGraw-Hill, New York.

HETSRONI, G. (ed.) (1982). *Handbook of multiphase systems.* McGraw-Hill, New York.

HEWITT, G. F. (1978). *Measurement of two-phase flow parameters.* Academic Press, London.

HEWITT, G. F. and HALL TAYLOR, N. S. (1971). *Annular two-phase flow.* Pergamon Press, Oxford.

WALLIS, G. B. (1969). *One-dimensional two-phase flow.* McGraw-Hill, New York.

2

TWO-PHASE FLOW PATTERNS AND FLOW PATTERN MAPS

2.1 Introduction

In gas–liquid flow the two phases can adopt various geometric configurations: these are known as flow patterns or flow regimes. Important physical parameters in determining the flow pattern are:

(*a*) surface tension, which keeps channel walls always wet (unless they are heated when they are usually wet) and which tends to make small liquid drops and small gas bubbles spherical; and

(*b*) gravity, which (in a non-vertical channel) tends to pull the liquid to the bottom of the channel.

The common flow patterns for vertical upflow, that is where both phases are flowing upwards, in a round tube are illustrated in Fig. 2.1. As the quality is gradually increased from zero, the flow patterns obtained are:

(*a*) bubbly flow, in which the gas (or vapour) bubbles are of approximately uniform size;

(*b*) plug flow (sometimes called slug flow), in which the gas flows as large bullet-shaped bubbles. (There are also some small gas bubbles distributed throughout the liquid.);

(*c*) churn flow, which is highly unstable flow of an oscillatory nature, for example the liquid near the tube wall continually pulses up and down;

(*d*) annular flow, in which the liquid travels partly as an annular film on the walls of the tube and partly as small drops distributed in the gas which flows in the centre of the tube.

The common flow patterns for horizontal flow in a round tube are illustrated in Fig. 2.2. As the quality is gradually increased from zero, the flow patterns obtained are:

(*a*) bubbly flow, in which the gas bubbles tend to flow along the top of the tube;

(*b*) plug flow, in which the individual small gas bubbles have coalesced to produce long plugs;

(*c*) stratified flow, in which the liquid–gas interface is smooth. Note that this flow pattern does not usually occur, the interface is almost

4

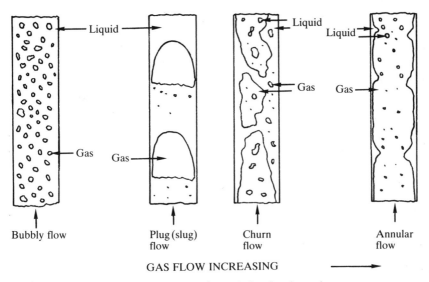

Fig. 2.1. Flow patterns in vertical upflow in a tube.

always wavy as in:

(*d*) wavy flow, in which the wave amplitude increases as the gas velocity increases;

(*e*) slug flow, in which the wave amplitude is so large that the wave touches the top of the tube; and

(*f*) annular flow, which is similar to vertical annular flow except that the liquid film is much thicker at the bottom of the tube than at the top.

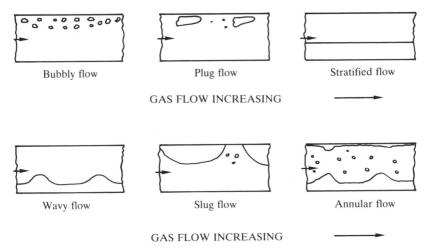

Fig. 2.2. Flow patterns in horizontal flow in a tube.

Many writers define other flow patterns, and nearly 100 different names have been used. Many of these are merely alternative names, while others delineate minor differences in the main flow patterns. The number of flow patterns shown in Figs. 2.1 and 2.2 probably represent the minimum which can sensibly be defined. Further general details about flow patterns can be found in Collier (1981) and Hewitt (1982).

2.2 Flow pattern determination

The main methods of determining the flow pattern are:

(a) by eye, but this is subjective and is only possible for flow in transparent tubes; and

(b) by various types of instrument, for example the gamma ray densitometer which measures mean density across the tube or average void fraction (the proportion of the total volume which is occupied by gas). The ideal response which might be obtained for various flow patterns might be as shown in Fig. 2.3, in which the flow patterns can easily be distinguished. However, the results are rarely so conclusive and so interpretation is again subjective. Such a method does, however, have the advantage that it can give results even for flow in an opaque tube. A much more detailed discussion of flow pattern determination is given by Hewitt (1978).

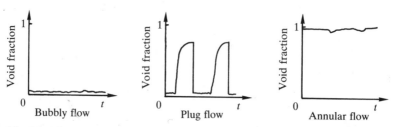

Fig. 2.3. Idealized response of a void-fraction probe to determine the flow pattern.

2.3 Transitions in vertical flow

It is interesting to look at the transition mechanisms at work as one flow pattern gives way to another. Here, only vertical flow is considered.

(a) *Bubbly to plug flow.* The normal mechanism for the transition is that, as the gas flow increases, the bubbles get closer together and collisions therefore occur more often (see Fig. 2.4, taken from Radovich and Moissis 1962). Some of the collisions lead to coalescence of bubbles and eventually to the formation of plugs. A large increase in collision

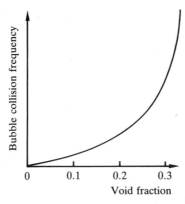

Fig. 2.4. Variation of bubble collision frequency with void fraction.

frequency at a void fraction of about 0.3 means that the transition tends to occur around this point. However, the transition may occur at a higher void fraction (possibly as high as 0.6) if coalescence is prevented by the presence of a surface active agent. Also, if the liquid flow rate (and therefore the liquid turbulence intensity) is high, large bubbles will be broken up into smaller ones even at high void fractions.

(b) *Plug to churn flow.* Because the gas plug rises through the liquid, the gas velocity in the plug is upwards, but the liquid velocity in the thin film around the plug is usually downwards, and so the flow is countercurrent (see Fig. 2.5). As will be seen in Chapter 11, at some definite flow rate the gas velocity will suddenly disrupt the liquid film (the film will 'flood') and Nicklin and Davidson (1962) therefore suggested that the plug flow will break down to give churn flow—a pulsating, highly unstable flow. Annular flow is not formed immediately because the gas velocity is not yet high enough to support the film continuously.

(c) *Churn to annular flow.* As the gas flow rate (and so the gas velocity) increases, the gas is able to support the liquid as a film on the

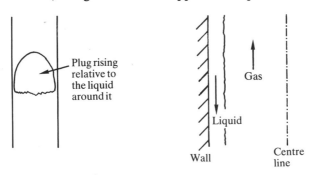

Fig. 2.5. Liquid and gas flows during the passage of a gas plug.

tube walls and annular flow occurs. This transition is related to the flow reversal point, which in turn is related to flooding. Flow reversal occurs when the gas velocity is reduced and the liquid first starts to flow down instead of up; again, this point is explored in more detail in Chapter 11 (see also Wallis 1961).

2.4 Flow pattern maps

Flow pattern maps are an attempt, on a two-dimensional graph, to separate the space into areas corresponding to the various flow patterns. Simple flow pattern maps use the same axes for all flow patterns and transitions. Complex maps use different axes for different transitions. Examples of some common, useful flow pattern maps are the following.

(a) The Baker map for horizontal flow (see Fig. 2.6). This map was first suggested by Baker (1954), and was subsequently modified by Scott (1963). The axes are defined in terms of G_g/λ and $G_l\psi$, where

$$G_g = \text{mass flux of gas (kg/m}^2\text{ s)} = \frac{\text{gas mass flow rate}}{\text{tube cross-sectional area}} \qquad (2.1)$$

$$G_l = \text{mass flux of liquid (kg/m}^2\text{ s)} = \frac{\text{liquid mass flow rate}}{\text{tube cross-sectional area}} \qquad (2.2)$$

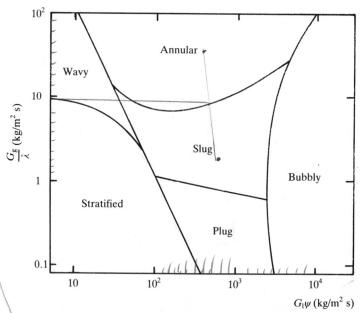

Fig. 2.6. Baker flow pattern map for horizontal flow in a tube.

$$\lambda = \left(\frac{\rho_g}{\rho_{air}} \frac{\rho_l}{\rho_{water}}\right)^{\frac{1}{2}} \tag{2.3}$$

$$\psi = \frac{\sigma_{water}}{\sigma}\left(\frac{\mu_l}{\mu_{water}}\left[\frac{\rho_{water}}{\rho_l}\right]^2\right)^{\frac{1}{3}} \tag{2.4}$$

Here: ρ_l is the liquid density (kg/m³); ρ_g is the gas density (kg/m³); ρ_{water} is the density of water = 1000 kg/m³; ρ_{air} is the density of air = 1.23 kg/m³; μ_l is the liquid viscosity (N s/m²); μ_{water} is the viscosity of water = 10^{-3} N s/m²; σ is the surface tension (N/m); and σ_{water} is the surface tension of air–water = 0.072 N/m.

The Baker map works reasonably well for water/air and oil/gas mixtures in small diameter (<0.05 m) tubes.

(b) The Hewitt and Roberts (1969) map for vertical upflow in a tube (see Fig. 2.7). Note that here G^2/ρ is a momentum flux, and so all the transitions are assumed to depend on the phase momentum fluxes. Wispy annular flow is a sub-category of annular flow which occurs at high mass flux when the entrained drops are said to appear as wisps or elongated droplets. The Hewitt and Roberts map works reasonably well for

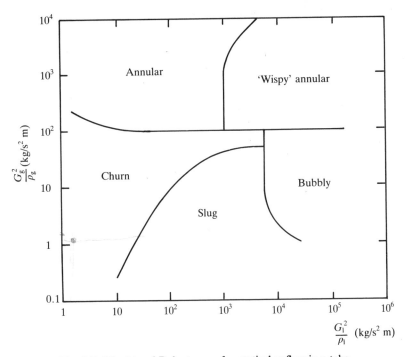

Fig. 2.7. Hewitt and Roberts map for vertical upflow in a tube.

water/air and water/steam systems over a range of pressures, again in small diameter tubes. The use of this map is explained in more detail in Example 1 in Chapter 25.

For both maps it should be noted also that the transitions between adjacent flow patterns do not occur suddenly but over a range of flow rates. Thus the lines should really be replaced by rather broad transition bands.

(c) The Taitel and Dukler (1976) map for horizontal flow. This is the best-known example of the complex type of flow pattern map. The map is shown in Fig. 2.8, and the parameters necessary are

$$X = \left[\frac{(dp/dz)_1}{(dp/dz)_g} \right]^{\frac{1}{2}}$$ (2.5)

$$Fr = \text{Froude number} = \frac{G_g}{[\rho_g(\rho_1 - \rho_g)\, dg]^{\frac{1}{2}}}$$ (2.6)

$$T = [|(dp/dz)_1|/g(\rho_1 - \rho_g)]^{\frac{1}{2}}$$ (2.7)

$$K = Fr \left[\frac{G_1 d}{\mu_1} \right]^{\frac{1}{2}}$$ (2.8)

Note that X is often known as the Martinelli parameter, and that the modulus sign in eqn (2.7) ensures that T is always positive.

$(dp/dz)_1 =$ the frictional pressure gradient if the liquid in the two-phase flow were flowing alone in the tube (N/m^3),

$(dp/dz)_g =$ the frictional pressure gradient if the gas in the two-phase flow were flowing alone in the tube (N/m^3),

$d =$ tube diameter (m),

$g =$ acceleration due to gravity $(9.81\ m/s^2)$, and

$\mu_1 =$ liquid viscosity $(N\ s/m^2)$.

Again, the transition lines should be shown as broad bands. All the transition criteria used by Taitel and Dukler have some theoretical basis, although this is sometimes rather tenuous. For example, the general approach is to take the flow rates of the phases and to work out the liquid depth h_1 (see Fig. 2.9) if the flow pattern were perfectly stratified. This is done by evaluating the pressure gradient in each phase and then adjusting h_1 to make the two pressure gradients equal. Taitel and Dukler say arbitrarily that intermittent flow will occur if $h_1/d > 0.5$, because a sinusoidal wave on the stratified flow will touch the top of the tube before the bottom. Such a crude approach is legitimately open to criticism but the overall method does, in general, produce sensible results capable of application to unusual fluids. The use of the map is explained in detail in Example 1 in Chapter 25. For a critical review of the method, see Hewitt (1982).

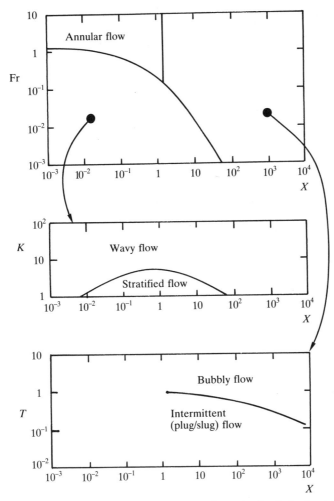

Fig. 2.8. Taitel and Dukler method for flow pattern determination in horizontal flow in a tube.

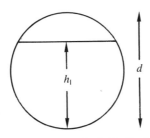

Fig. 2.9. Taitel and Dukler method: idealized stratified flow.

A complex type of flow pattern map has also been proposed for vertical flow but it has not been so well tested.

2.5 Effect of heat flux

If the flow is heated, as with water evaporating in a boiler tube, then the heat flux does have some effect on the flow patterns (see Fig. 2.10). Note that the quality used here is the thermodynamic quality, which is defined by

$$x = \frac{h - h_{1,\text{sat}}}{\lambda},\qquad (2.9)$$

where: h is the mixture enthalpy (J/kg); $h_{1,\text{sat}}$ is the liquid saturation enthalpy (J/kg); and λ is the enthalpy of vaporization (J/kg).

From Fig. 2.10 it can be noted that bubbly flow occurs before thermodynamic quality reaches zero. Also, the flow pattern called 'drop flow' is found. This consists of a flow of liquid drops with a dry tube wall. Eventually all the drops evaporate, but only when the quality exceeds unity. Thus, the effect of heat flux on the flow pattern is twofold. First, it causes flow patterns to occur at lower qualities than in adiabatic flow and second, it causes new flow patterns to occur.

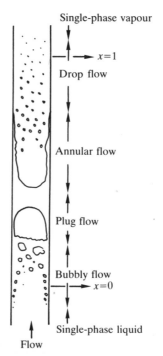

Fig. 2.10. Flow patterns in an evaporating flow.

2.6 Other geometries

Flow patterns have been studied in other geometries and generally behave in ways rather similar to the round tube behaviour.

(a) *Rectangular channels.* The flow patterns are very similar to those in round tubes, however liquid tends to collect in the corners.

(b) *Vertical downflow* (*in tubes*). When both phases are flowing downwards annular flow occurs very frequently (see Barnea *et al.* 1980; Golan and Stenning 1969).

(c) *Obstructions on the wall.* There is a tendency for the liquid to be thrown into the centre of the channel.

(d) *Helical twisted tape insert.* This tends to throw the liquid on to the walls of the tube (so producing good heat transfer).

(e) *Helically coiled tube* (*see Fig. 2.11*). Again, this tends to throw liquid on to the walls of the tube at point A or point B or somewhere in between (on ACB) due to the effect of gravity. This is discussed further in Chapter 13 (see also Banerjee *et al.* 1969).

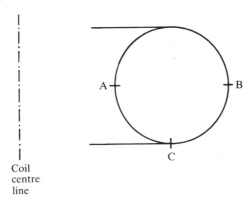

Coil
centre
line

Fig. 2.11. Helical coil: cross sectional view.

(f) *Rod bundles* (*as in nuclear reactors*). The flow patterns are again very similar to those found in round tubes, though of course with many detailed differences. Round-tube flow pattern maps even give reasonable results when used for axial flow along a bundle of tubes (see Bergles 1969).

References

BAKER, O. (1954). Simultaneous flow of oil and gas. *Oil Gas J.* **53,** 185.

BANERJEE, S., RHODES, E., and SCOTT, D. S. (1969). Studies on cocurrent gas–liquid flow in helically coiled tubes: flow patterns, pressure drop, and hold-up. *Can. J. Chem. Engnrs* **47,** 445–517.

BARNEA, D., SHOHAN, O., TAITEL, Y., and DUKLER, A. E. (1980). Flow pattern transition for horizontal and inclined pipes: experimental and comparison with theory. *Int. J. Multiphase Flow* **6**, 217–25.

BERGLES, A. E. (1969). Two-phase flow structure observations for high pressure water in a rod bundle. *Proc. ASME Symp. Two-Phase Flow Heat Transfer in Rod Bundles*, 47–55.

COLLIER, J. G. (1981). *Convective boiling and condensation* (2nd ed), pp. 8–23. McGraw-Hill, New York.

GOLAN, P. L. and STENNING, A. H. (1969). Two-phase vertical flow maps. *Proc. Instn. Mech. Engnrs* **184(3C)**, 110–16.

HEWITT, G. F. (1978). *Measurement of two-phase flow parameters*, pp. 105–11. Academic Press, London.

HEWITT, G. F. (1982). Flow regimes. In *Handbook of multiphase systems* (ed. G. Hetsroni), pp. 2-3–2-43, McGraw-Hill, New York.

HEWITT, G. F. and ROBERTS, D. N. (1969). Studies of two-phase flow patterns by simultaneous flash and X-ray photography. *AERE-M2159*.

NICKLIN, D. J. and DAVIDSON, J. F. (1962). The onset of instability in two-phase slug flow. *Instn of Mech. Engnrs Proc. Symp. on Two-Phase Flow*, paper 4.

RADOVICH, N. A. and MOISSIS, R. (1962). The transition from two-phase bubble flow to slug flow. *MIT Report 7-7673-22*.

SCOTT, D. S. (1963). Properties of co-current gas–liquid flow. In *Advances in chemical engineering*, **4**, 199–277.

TAITEL, Y. and DUKLER, A. E. (1976). A model for predicting flow regime transitions in horizontal and near horizontal gas–liquid flow. *AIChE J.* **22**, 47–55.

WALLIS, G. B. (1961). Flooding velocities for air and water tubes. *AEEW-R123*.

3

BUBBLY AND PLUG FLOW

3.1 Behaviour of single, very small bubbles

If bubbles are very small, surface tension makes them spherical. Then, if the Reynolds number is low enough, the drag on the bubble can be calculated from Stokes' law. The criterion for Stokes' law to apply is that the Reynolds number $Re\,(=\rho_l Du_b/\mu_l)$ should be much less than unity. The drag force is then equal to $3\pi Du_b\mu_l$. Here: ρ_l is the liquid density (kg/m^3); μ_l is the liquid viscosity (N s/m^2); D is the bubble diameter (m); and u_b is the bubble velocity (m/s) through stagnant liquid.

In terms of the drag coefficient C_D

$$C_D = \frac{\text{drag force}}{\text{projected area} \times \text{dynamic head}}$$

$$= \frac{3\pi Du_b\mu_l}{\frac{1}{4}\pi D^2 \times \frac{1}{2}\rho_l u_b^2} = \frac{24}{Re}. \tag{3.1}$$

Figure 3.1 shows how the drag coefficient actually varies with Reynolds number; it can be seen that Stokes' law actually works well up to a Reynolds number of about unity. However, even a Reynolds number of unity represents a small bubble: for air–water under atmospheric conditions the corresponding bubble diameter is about 0.1 mm.

In order to find the terminal rising velocity the drag force and the buoyancy force are equated.

$$3\pi Du_b\mu_l = \frac{\pi D^3}{6}(\rho_l - \rho_g)g \tag{3.2}$$

and therefore

$$u_b = \frac{D^2 g(\rho_l - \rho_g)}{18\mu_l}. \tag{3.3}$$

So far the bubble has been treated like a solid particle which is lighter than the surrounding fluid. The bubble is, however, gaseous and there is apparently no reason why the interface should actually be stationary. If the interface can move then the gas inside the bubble will also move: see Fig. 3.2 which shows the streamlines in a bubble held stationary by a downflow of liquid. The full equations of motion—the Navier–Stokes equations—can be solved. This was first done independently by

15

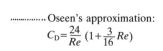

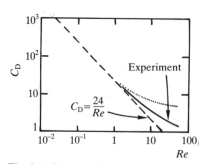

Fig. 3.1. Drag coefficient for a spherical particle.

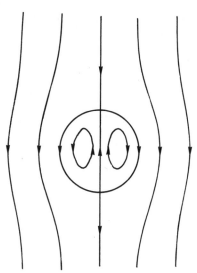

Fig. 3.2. Streamlines in and around a bubble with a mobile surface.

Hadamard and Rybczynski. The solution, for u_b (see Batchelor 1967), is

$$u_b = \frac{D^2 g(\rho_1 - \rho_g)}{18\mu_1} \left(\frac{3\mu_g + 3\mu_1}{3\mu_g + 2\mu_1} \right), \tag{3.4}$$

where μ_g = gas viscosity (N s/m^2).
For the usual case, where $\mu_1 \gg \mu_g$, eqn (3.4) reduces to

$$u_b = \frac{D^2 g(\rho_1 - \rho_g)}{12\mu_1}. \tag{3.5}$$

Thus the effect of allowing the interface to move is that u_b has been increased by 50 %. However, most experiments show that the Stokes' law solution (eqn (3.3)) works better. This is almost certainly because any surface active agent in the system will render the surface immobile. Surface active molecules migrate to the surface and get swept along by the moving surface. They thus collect at the rear of the bubble and the resulting surface tension gradient prevents any further movement of the surface.

3.2 Behaviour of intermediately sized bubbles

Larger bubbles are not spherical, nor do they obey Stokes' law. A generalized chart giving information about their shape and rising velocity is shown in Fig. 3.3 (from Clift et al. 1978).

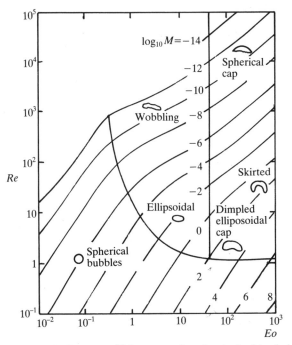

Fig. 3.3. Reynolds, Eotvos and Morton numbers for single rising bubbles.

In this case

$$Eo = \text{Eotvos number} = \frac{g(\rho_1 - \rho_g)D_e^2}{\sigma} \qquad (3.6)$$

$$Re = \text{Reynolds number} = \frac{\rho_1 D_e u_b}{\mu_1} \qquad (3.7)$$

$$M = \text{Morton number} = \frac{g\mu_1^4(\rho_1 - \rho_g)}{\rho_1^2 \sigma^3}, \qquad (3.8)$$

where: D_e is the equivalent diameter (m) (the diameter the bubble would have if it were spherical); and σ is the surface tension (N/m).
The rising velocity of a bubble can therefore be found by

(a) calculating the Morton number from eqn (3.8),
(b) calculating the Eotvos number from eqn (3.6),
(c) looking up the Reynolds number from Fig. 3.3, and then
(d) the rising velocity u_b can be found from eqn (3.7).

Further details are given in Example 2 in Chapter 25.

For air–water at 1 bar, $M = 3 \times 10^{-11}$ so the shape transitions are

Spherical ⟶ Wobbling ⟶ Spherical cap

Transition at

$Eo = 0.35$	$Eo = 40$
$Re = 500$	$Re = 3000$
$D_e = 1.6\,\text{mm}$	$D_e = 17\,\text{mm}$
$u_b = 0.3\,\text{m/s}$	$u_b = 0.2\,\text{m/s}$

Thus, as diameter increases by a factor of 10, the rising velocity decreases. This is because the shape and the drag force are changing. The rising velocity characteristics of air bubbles in water are shown in Fig. 3.4.

Spherical cap bubbles are bubbles shaped like the top of a sphere (see Fig. 3.5). For $Re > 150$, the half angle θ subtended at the centre of the sphere is approximately $50°$, and base is flat and irregular. At lower values of the Reynolds number the angle θ increases and the wake becomes closed instead of open.

If the observed spherical cap shape is assumed (see Fig. 3.6), the rising velocity can be calculated. This was first done by Davies and Taylor (see Davidson and Harrison 1963). If the bubble is held stationary by a downflow of liquid with velocity u_b, then from Bernoulli's equation in the liquid

$$p_\theta - p_0 = \rho_l g R(1 - \cos \theta) - \tfrac{1}{2}\rho_l u_\theta^2 \qquad (3.9)$$

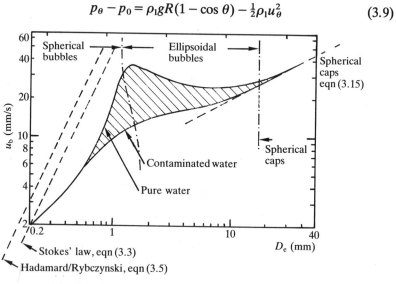

Fig. 3.4. Rising velocity of single air bubbles in water.

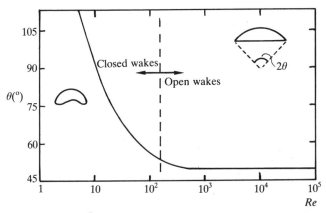

Fig. 3.5. Spherical cap bubbles: influence of Reynolds number on wake angle.

where u is the velocity of liquid (m/s), p is the pressure (N/m^2) (see Fig. 3.6), and R is the radius of curvature (m).

The inviscid solution for flow around a sphere works well over the front half of the sphere. It is

$$u_\theta = \tfrac{3}{2}u_b \sin \theta. \tag{3.10}$$

In the gas, if the gas is stagnant,

$$p_\theta - p_0 = \rho_g g R(1 - \cos \theta). \tag{3.11}$$

Hence equating $p_\theta - p_0$

$$u_b^2 = \tfrac{8}{9}gR\left(\frac{\rho_1 - \rho_g}{\rho_1}\right)\left[\frac{1 - \cos \theta}{\sin^2\theta}\right]. \tag{3.12}$$

The bubble rising velocity cannot depend on θ, but $f(\theta) = (1 - \cos \theta)/\sin^2\theta$ is a weak function of θ over the range 0 to 50°, and

$$\underset{\theta \to 0}{\text{Limit}} \frac{1 - \cos \theta}{\sin^2\theta} = \frac{1}{2}. \tag{3.13}$$

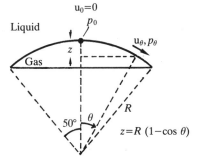

Fig. 3.6. Spherical cap bubble.

Actually, the shape of the bubble is not quite spherical, and the shape is such that $f(\theta) = \text{constant}$. Taking the limiting value of $f(\theta)$ given by eqn (3.13)

$$u_b = \frac{2}{3}\left[gR\left(\frac{\rho_1 - \rho_g}{\rho_1}\right)\right]^{\frac{1}{2}},$$ (3.14)

and then using geometry to relate R to D_e

$$u_b = 0.71\left[gD_e\left(\frac{\rho_1 - \rho_g}{\rho_1}\right)\right]^{\frac{1}{2}}.$$ (3.15)

This is the spherical cap equation used in Fig. 3.4.

For air–water, it can be noted that as D_e changes from 2 mm to 20 mm (a volume change of a factor of 1000) the rising velocity only changes by about 20 %. The change in shape from spherical to wobbling to spherical cap continually increases the drag, and so the expected large increase in rising velocity does not occur.

3.3 Behaviour of individual plugs

As the volume of a spherical cap bubble is increased, when D_e/d (where d = tube diameter) reaches values of greater than about 0.6, the bubble motion is controlled not by the equivalent diameter D_e but by the tube diameter d. The bubble has now become a plug with its characteristic bullet-shaped nose (see Fig. 3.7). The rising velocity can be expressed as a Froude number

$$Fr = \left[\frac{\rho_1}{\rho_1 - \rho_g}\right]^{\frac{1}{2}}\frac{u_p}{(gd)^{\frac{1}{2}}},$$ (3.16)

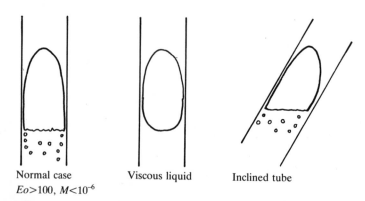

Normal case   Viscous liquid   Inclined tube
$Eo > 100, M < 10^{-6}$

Fig. 3.7. Plug shapes.

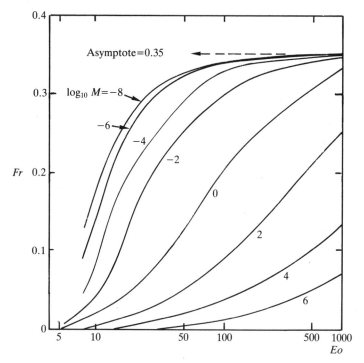

Fig. 3.8. Froude number variation with Eotvos and Morton number for a single plug.

where u_p is the plug rising velocity (m/s). For many cases, when $Eo > 100$ and $M < 10^{-6}$, then $Fr = 0.35$. Note that in the definition of the Eotvos number (see eqn (3.6)) d is used in place of D_e. A more complete representation is given by Fig. 3.8 (from Wallis 1969). This complete representation is used in Example 3 in Chapter 25.

For low-pressure gas plugs in a reasonably sized tube (diameter greater than about 10 mm) and a low viscosity liquid, u_p is given by the simple equation

$$u_p = 0.35(gd)^{\frac{1}{2}}. \tag{3.17}$$

It can be noted that in this equation:

(a) fluid physical properties do not occur,
(b) u_p is only dependent on the tube diameter, and
(c) u_p is not dependent upon the length of the plug; this independence breaks down when the plug length is less than $1.5\,d$.

It might be expected that in an inclined tube a plug would rise more slowly, but this is not always true (see Fig. 3.9). The plug velocity often reaches a maximum at an inclination of about 45° (Zukoski 1966). Not

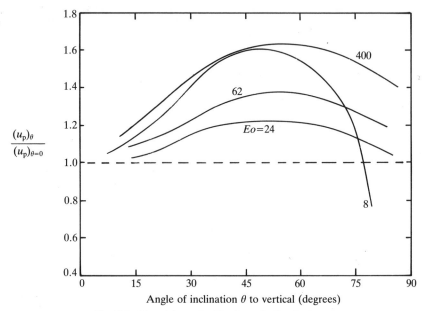

$$\frac{(u_{\mathrm{p}})_\theta}{(u_{\mathrm{p}})_{\theta=0}}$$

Angle of inclination θ to vertical (degrees)

Fig. 3.9. Plug rising velocities in an inclined tube.

only is the velocity along the tube direction increased, so also is the vertical component of the velocity. Part of the explanation for this increase in rising velocity in inclined tubes is the changed shape of the plug, particularly at the nose.

3.4 Slip velocity

If one bubble is rising through stagnant liquid then the velocity of the bubble relative to the liquid (the *slip velocity*) is equal to u_{b}. It can be argued that, as far as the bubble is concerned (as long as there is no acceleration), then the slip velocity is the only important velocity. If the liquid were moving upwards at a velocity of u_1, then the bubble rising velocity would be $u_1 + u_{\mathrm{b}}$ (giving a slip velocity of u_{b} again).

In a cloud of bubbles with void fraction α, a number of suggestions about the slip velocity u_{s} have been made. Two of the simplest and best are those of:

(*a*) Turner (1966)

$$u_{\mathrm{s}} = u_{\mathrm{b}} \tag{3.18}$$

that is, u_{s} is unaffected by α; and

(*b*) Wallis (1969)

$$u_{\mathrm{s}} = u_{\mathrm{b}}(1 - \alpha). \tag{3.19}$$

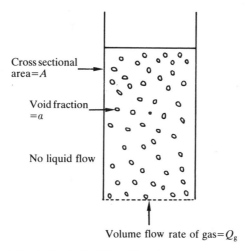

Fig. 3.10. A cloud of bubbles rising through stagnant liquid.

If we now consider the situation shown in Fig. 3.10, where gas is being bubbled through a stagnant liquid, then the superficial gas velocity (gas velocity if no liquid were present) is given by

$$V_g = Q_g/A. \tag{3.20}$$

The actual velocity of gas bubbles $= V_g/\alpha$, and so the slip velocity $u_s = V_g/\alpha$, since the liquid velocity is zero. Then

(*a*) using eqn (3.18) gives

$$V_g = \alpha u_b \tag{3.21}$$

(*b*) and using eqn (3.19) gives

$$V_g = \alpha(1 - \alpha)u_b. \tag{3.22}$$

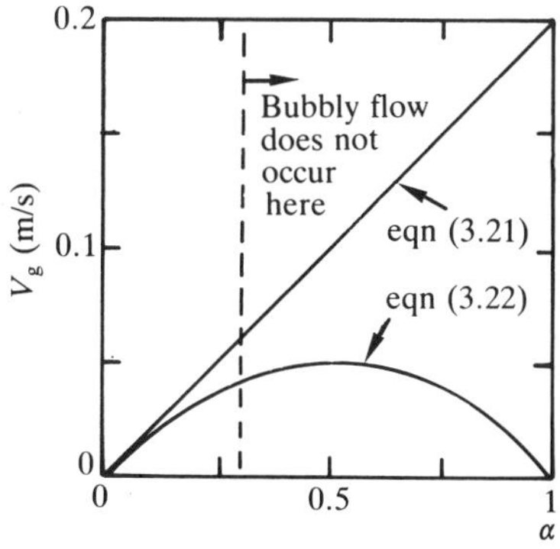

Fig. 3.11. Bubbly flow: dependence of void fraction on the superficial gas velocity.

For a typical value of u_b of 0.2 m/s the curves may seem very different, but bubbly flow generally only occurs when the void fraction is less than about 0.3. In this region it is not easy, experimentally, to distinguish the two curves in Fig. 3.11.

3.5 Slip velocity and plug flow

Here we are concerned with the physical meaning of u_p in plug flow. One possible experiment which helps to define u_p is shown in Fig. 3.12. In this experiment u_p is the velocity of the plug up the tube. Clearly, in this case, the velocity of the gas is u_p and the velocity of liquid above the plug is zero, hence the slip velocity is u_p. Therefore, u_p is actually the slip velocity in plug flow.

Now consider a more normal experiment where gas is pumped into a tube containing liquid, but where there is no net liquid flow, as in Fig. 3.13.

In this case

$$\text{gas velocity} - \text{liquid velocity} = \text{slip velocity.} \qquad (3.23)$$

As shown in Fig. 3.13, the control volume enables us to see that the liquid velocity across a plane not containing a gas plug is equal to V_g, so

$$\text{gas velocity} = u_p + V_g. \qquad (3.24)$$

But this gas velocity is the actual velocity of the plugs and is also equal to V_g/α. Hence

$$V_g = \frac{\alpha}{1-\alpha} u_p = \frac{\alpha}{1-\alpha} 0.35(gd)^{\frac{1}{2}}. \qquad (3.25)$$

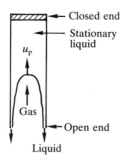

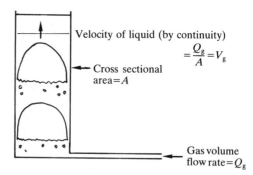

Fig. 3.12. Tube emptying experiment to determine the plug rising velocity.

Fig. 3.13. Steady plug flow.

If now we recognize that the liquid centre-line velocity is the important velocity as far as the plug is concerned, and that this is, in turbulent flow, about 20 % greater than the mean velocity, then the above equations become

$$\text{gas velocity} = u_p + 1.2V_g \tag{3.26}$$

and

$$V_g = \frac{\alpha}{1 - 1.2\alpha} u_p = \frac{\alpha}{1 - 1.2\alpha} 0.35(gd)^{\frac{1}{2}}. \tag{3.27}$$

These points will be explored in more detail when the drift flux model of two-phase flow is considered in Chapter 7. Nicklin *et al.* (1962) point out that a range of possible experiments to determine the plug rising velocity are possible, and that their interpretation is not straightforward.

References

BATCHELOR, G. K. (1967). *An introduction to fluid dynamics*. Cambridge University Press.

CLIFT, R., GRACE, J. R., and WEBER, M. E. (1978). *Bubbles, drops and particles*. Academic Press, New York.

DAVIDSON, J. F. and HARRISON, D. (1963). *Fluidised particles*. Cambridge University Press.

NICKLIN, D. J., WILKES, J. O., and DAVIDSON, J. F. (1962). Two-phase flow in vertical tubes. *Trans. Instn Chem. Engnrs* **40**, 61–8.

TURNER, J. C. R. (1966). On bubble flow in liquids and fluidised beds. *Chem. Eng. Sci.* **21**, 971–4.

WALLIS, G. B. (1969). *One dimensional two-phase flow*. McGraw-Hill, New York.

ZUKOSKI, E. E. (1966). The motion of long bubbles in closed tubes. *J. Fluid Mech.* **25**, 821–37.

4

ANNULAR FLOW

4.1 Introduction

The main features of annular flow are shown in Fig. 4.1. Here only vertically upwards annular flow is considered; both the liquid and the gas phases are travelling upwards. The liquid flows partly as a thin liquid film on the channel walls and partly as drops entrained into the gas core. Annular flow occurs over a wide range of qualities. At low pressures (approximately 1 bar) annular flow is found when the quality is only a few per cent.

The liquid film is never smooth but always has ripples on its surface, and usually has large disturbance waves. These waves are described in detail in Chapter 9. The waves are the main source of entrainment of liquid, in the form of drops, into the gas core. The processes of entrainment and deposition (of drops back on to the liquid film) will also be discussed in Chapter 9. Many details, experimental and theoretical, of annular flow are discussed by Hewitt and Hall Taylor (1970).

For the moment, it will be assumed that all the liquid is travelling in the film. This is sometimes a reasonable assumption, but in extreme cases up to 80 % or 90 % of the liquid may be flowing as drops. In such cases, of course, the assumption is very unrealistic.

4.2 Liquid film behaviour—shear stress variation

A force balance on an annular ring of outer radius r, inner radius r_i, and axial length δz (see Fig. 4.2), assuming the flow is steady so there are no acceleration terms, gives

$$\underset{\substack{\text{shear force on} \\ \text{outer surface}}}{-2\pi r \tau \delta z} + \underset{\substack{\text{shear force on} \\ \text{inner surface}}}{2\pi r_i \tau_i \delta z} - \underset{\substack{\text{gravitational force}}}{\rho_l g \delta z \pi (r^2 - r_i^2)}$$

$$+ \underset{\text{pressure force}}{\left[p - \left(p + \frac{\mathrm{d}p}{\mathrm{d}z} \delta z \right) \right] \pi (r^2 - r_i^2)} = 0, \qquad (4.1)$$

where:

   τ = shear stress at radius r (N/m²), and
   τ_i = interfacial shear stress (at r_i) (N/m²).

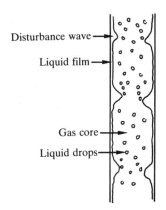

Fig. 4.1. Main features of annular flow.

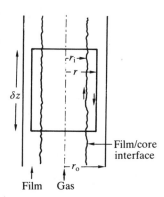

Fig. 4.2. Element of the flow for force balance.

Hence

$$\tau = \tau_i \frac{r_i}{r} - \frac{1}{2}\left(\rho_l g + \frac{dp}{dz}\right)\left(\frac{r^2 - r_i^2}{r}\right). \tag{4.2}$$

The interfacial shear stress τ_i and the pressure gradient dp/dz can be related by a force balance on a cylinder, radius r, entirely in the gas core

$$\tau = -\frac{r}{2}\left(\rho_g g + \frac{dp}{dz}\right) \tag{4.3}$$

and hence, at r_i,

$$\tau_i = -\frac{r_i}{2}\left(\rho_g g + \frac{dp}{dz}\right). \tag{4.4}$$

It should be remembered that dp/dz is necessarily negative (the pressure falls with increasing axial length), and that $(\rho_g g + dp/dz)$ is, overall, still a negative quantity.

The most usual variation of shear stress across the entire cross section is shown in Fig. 4.3. If the liquid film is thin, then it can often be treated as a constant shear stress film.

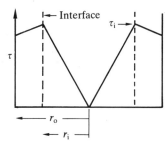

Fig. 4.3. Typical shear stress variation across the tube.

If the shear stress variation in the film is known, then in principle the velocity profile in the film can be found. The cases of laminar flow and turbulent flow in the liquid film are both now considered.

4.3 Liquid film behaviour—velocity profile

The simplest assumption is that the liquid film is laminar, in which case

$$\tau = \mu_1 \frac{du}{dy}, \tag{4.5}$$

where:

u = velocity in film (m/s), and
y = distance from the wall (m).

Obviously with this assumption the full equation for shear stress variation in the film can be integrated to give the velocity variation

$$u = \frac{1}{\mu_1}\left\{\left[\tau_i r_i + \frac{1}{2}\left(\rho_1 g + \frac{dp}{dz}\right)r_i^2\right]_1 \ln\frac{r_o}{r} - \frac{1}{4}\left(\rho_1 g + \frac{dp}{dz}\right)(r_o^2 - r^2)\right\}. \tag{4.6}$$

If the shear stress is constant (and equal to τ), then the equation for u is much simpler

$$u = \frac{\tau y}{\mu_1} = \frac{\tau}{\mu_1}(r_o - r) \tag{4.7}$$

The film is usually turbulent, however, and so these equations do not usually apply. If we assume that the film behaves like the boundary layer region in single-phase pipe flow then we can use results from single-phase theory. For example, the universal velocity profile can be used

$$u^+ = f(y^+), \tag{4.8}$$

where

u^+ = dimensionless velocity = u/u^*,
u^* = friction velocity (m/s) = $(\tau/\rho_1)^{\frac{1}{2}}$, and
y^+ = dimensionless distance from wall = $yu^*\rho_1/\mu_1$

The universal velocity profile is a three-part equation in which the function, eqn (4.8), depends on the value of y^+

laminar sub-layer: $u^+ = y^+$ for $y^+ < 5$,   (4.9)

buffer layer: $u^+ = -3.05 + 5 \ln y^+$ for $5 < y^+ < 30$,   (4.10)

turbulent core: $u^+ = 5.5 + 2.5 \ln y^+$ for $y^+ > 30$.   (4.11)

Taking the laminar sub-layer (eqn (4.9)) $u^+ = y^+$ and substituting for u^+ and y^+

$$u\left(\frac{\rho}{\tau}\right)^{\frac{1}{2}} = y\left(\frac{\tau}{\rho_1}\right)^{\frac{1}{2}}\left(\frac{\rho_1}{\mu_1}\right),\qquad(4.12)$$

or

$$u = \frac{\tau y}{\mu_1}\qquad(4.7)$$

This is the same result as was previously obtained for a laminar film and constant shear stress.

4.4 Integration of the velocity profile—triangular relationship

If the film is so thin that it can be assumed to be flat, then curvature effects can be neglected. This can be done if $m/d \ll 1$, where $m =$ film thickness (m), and $d =$ tube diameter (m). Then the liquid film flow rate can be found by integrating the velocity profile across the liquid film

$$W_{lf} = \pi d \int_0^m \rho_1 u \, dy,$$

where $W_{lf} =$ liquid film flow rate (kg/s). For the simplest laminar flow case, $u = \tau y/\rho_1$, so

$$W_{lf} = \pi d \rho_1 \frac{\tau}{\mu_1} \int_0^m y \, dy = \pi d\tau \frac{\rho_1}{2\mu_1} m^2 \qquad(4.13)$$

For the universal velocity profile in turbulent flow

$$W_{lf} = \pi d \rho_1 \int_0^{m^+} u^+ u^* \frac{\mu_1}{\rho_1 u^*} \, dy^+ \qquad(4.14)$$

or, simplifying,

$$W_{lf} = \pi d \mu_1 \int_0^{m^+} u^+ \, dy^+ \qquad(4.15)$$

This last equation suggests the use of a dimensionless liquid film flow rate $W_{lf}^+ = W_{lf}/\pi d\mu_1$, and then

$$W_{lf}^+ = \int_0^{m^+} u^+ \, dy^+ \qquad(4.16)$$

Now, integrating eqns (4.9–11) gives

$$W_{lf}^+ = 0.5m^{+2} \qquad\qquad\qquad \text{for} \quad m^+ < 5 \qquad (4.17)$$

$$W_{lf}^+ = 12.5 - 8.05m^+ + 5m^+\ln m^+ \quad \text{for} \quad 5 < m^+ < 30 \qquad (4.18)$$

$$W_{lf}^+ = -64 + 3m^+ + 2.5m^+\ln m^+ \quad \text{for} \quad m^+ > 30 \qquad (4.19)$$

Here the solutions have been matched at the changeover points so, for example, eqns (4.18) and (4.19) give the same value of W_{lf}^+ for $m^+ = 30$. Equations like these are often called the *triangular relationship* because they relate three variables: liquid film flow rate W_{lf}, film shear stress τ, and average film thickness m.

Take a particular example of a flow in which

tube diameter $d = 0.032$ m,
liquid film flow rate $W_{lf} = 0.037$ kg/s,
liquid density $\rho_1 = 1000$ kg/m^3,
liquid viscosity $\mu_1 = 10^{-3}$ Ns/m^2, and
pressure gradient $-\mathrm{d}p/\mathrm{d}z = 2100$ N/m^3.

Then the shear stress τ is given approximately by

$$\tau = -\frac{\mathrm{d}p}{\mathrm{d}z}\frac{d}{4} = 16.8\,\text{N/m}^2, \qquad (4.20)$$

and so

$$W_{lf}^+ = \frac{W_{lf}}{\pi d \mu_1} = 368. \qquad (4.21)$$

For $W_{lf}^+ = 368$, by trial and error, the universal velocity profile equations (integrated) give $m^+ = 36.1$ (here eqn (4.19) is the relevant one). The actual film thickness m can then be found from

$$m = m^+ \frac{\mu_1}{\rho_1 u^*} = m^+ \frac{\mu_1}{(\tau\rho_1)^{\frac{1}{2}}}. \qquad (4.22)$$

In this case

$$m = \frac{36.1 \times 10^{-3}}{(16.8 \times 10^3)^{\frac{1}{2}}} = 279 \times 10^{-6}\,\text{m} = 279\,\mu\text{m}, \qquad (4.23)$$

and the actual value determined by experiment was 315 μm.

Generally, the triangular relation seems to give answers to within about 20 % (see, for example, Fig. 4.4). This is remarkable considering that

(a) the velocity profiles were developed for single-phase flow, and
(b) the film is not smooth but carries large waves.

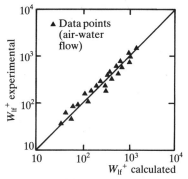

Fig. 4.4. Triangular relation: typical calculated and experimental results for film flow rate (from Gill *et al.* 1962).

4.5 Interfacial roughness correlation

If we know the liquid film flow rate and we are interested in calculating the average film thickness and the shear stress (and hence the pressure gradient), the triangular relationship gives one equation—another is therefore needed. This is the interfacial roughness correlation which arose first from measurements of velocity profiles in the gas (see Fig. 4.5).

As the liquid flow rate was increased and the gas flow rate held constant (and so the film thickness increased) the gas velocity profile became more peaked. It looked almost like the parabolic velocity profile of a laminar flow. In fact the velocity profile becomes more peaked as a result of the increasing roughness presented by the thick liquid film (with its large disturbance waves) to the gas flow. A similar phenomenon occurs in single phase flow in rough walled tubes.

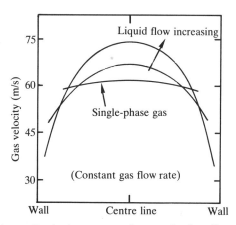

Fig. 4.5. Velocity profiles in the gas core of an annular flow (from Gill *et al.* 1964).

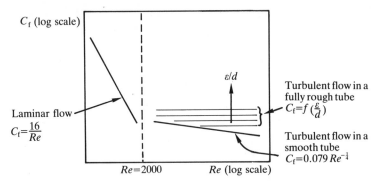

Fig. 4.6. Friction factors in single-phase flow.

Single-phase results give friction factor (C_f) as a function of Reynolds number (Re) and roughness (ε) of the tube, as shown schematically in Fig. 4.6. In general,

$$C_f = f(Re, \ \varepsilon/d).\tag{4.24}$$

Treating the gas flow as a single phase flow, the friction factor and Reynolds number for the gas can be calculated

$$C_f = \frac{\tau}{\frac{1}{2}\rho_g u_g^2} \approx \frac{(-dp/dz)d\rho_g}{2G_g^2},\tag{4.25}$$

where u_g = gas core velocity (m/s). This is approximately equal to the gas superficial velocity, and G_g is the gas mass flux (kg/m^2 s). The appropriate Reynolds number for use in eqn (4.24) is then the gas Reynolds number Re_g

$$Re_g = \frac{dG_g}{\mu_g},\tag{4.26}$$

where μ_g = gas viscosity (N s/m^2). Hence ε/d can be found from Fig. 4.6. ε was found to be a unique function of the film thickness m (see Fig. 4.7).

This type of roughness correlation has not been developed very far. Interest has mainly centred on another, simpler, way of expressing the same idea, which is due to Wallis (1970). An interfacial friction factor C_{fi} was defined by

$$C_{fi} = \frac{\tau_i}{\frac{1}{2}\rho_g u_g^2},\tag{4.27}$$

and then this friction factor was assumed to be given by

$$C_{fi} = C_{fg}\left(1 + 360\frac{m}{d}\right).\tag{4.28}$$

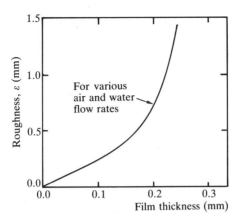

Fig. 4.7. Variation of apparent film roughness ε with liquid film thickness (from Gill *et al.* 1963).

C_{fg} is the gas single-phase friction factor in a smooth tube, which is given by

$$C_{fg} = 0.079\, Re_g^{-\frac{1}{4}} \qquad (4.29)$$

The procedure, if the film thickness is known and the shear stress is to be calculated, is therefore:

  (a) calculate Re_g from eqn (4.26);
  (b) calculate C_{fg} from eqn (4.29);
  (c) calculate C_{fi} from eqn (4.28); and then
  (d) calculate τ_i from eqn (4.27).

Once the shear stress is known, then the pressure gradient can be estimated.

4.6 Simultaneous solution of the triangular relationship and the interfacial roughness correlation

It is now clear that the triangular relationship and the interfacial roughness correlation are both equations (or sets of equations) connecting the average film thickness and the shear stress. They can be solved simultaneously for these unknowns, but the solution is not trivial since both equations are not linear. Trial and error is one obvious method of solution.

As stated before, the triangular relationship works very well, and typical errors are in the region of 20 %. The interfacial roughness correlation is much less well tested and the errors are certainly greater. For high-pressure systems it seems that particularly large errors occur. It has been suggested (Whalley and Hewitt 1978) that the equation

$$C_{fi} = C_{fg}\left(1 + 360\frac{m}{d}\right) \tag{4.28}$$

should be replaced by

$$C_{fi} = C_{fg}\left[1 + 24\left(\frac{\rho_l}{\rho_g}\right)^{\frac{1}{3}}\frac{m}{d}\right], \tag{4.30}$$

although the evidence for this is far from clear cut. The overall accuracy of calculations for shear stress and film thickness is discussed in more detail in Chapter 10.

References

GILL, L. E., HEWITT, G. F., and LACEY, P. M. C. (1962). Data on the upwards annular flow of air–water mixtures. *Chem. Eng. Sci.* **17**, 71–88.

GILL, L. E., HEWITT, G. F., and LACEY, P. M. C. (1964). Sampling probe studies of the gas core in annular two-phase flow. II Studies of the effect of phase flowrates on phase and velocity distribution. *Chem. Eng. Sci.* **19**, 665–82.

GILL, L. E., HEWITT, G. F., HITCHON, J. W., and LACEY, P. M. C. (1963). Sampling probe studies of the gas core distribution in annular two-phase flow. I The effect of length on phase and velocity distribution. *Chem. Eng. Sci.* **18**, 525–35.

HEWITT, G. F. and HALL Taylor, N. S. (1970). *Annular two-phase flow.* Pergamon Press, Oxford.

WALLIS, G. B. (1970). Annular two-phase flow. II Additional effects. *J. Basic Engineering* **92**, 73–82.

WHALLEY, P. B., and HEWITT, G. F. (1978). The correlation of liquid entrainment fraction and entrainment rate in annular two-phase flow. *AERE-R9187.*

5

HOMOGENEOUS FLOW

5.1 Introduction

Homogeneous flow is a particular model or picture of two-phase flow. In this model it is assumed that the two phases are well mixed and, more particularly, travelling with equal velocities.

We will first examine the continuity equations (see Fig. 5.1). Now remembering that

$$G = \text{overall mass flux (kg/m}^2\,\text{s)}$$
$$= \text{total mass flow rate/total cross-sectional area,} \qquad (5.1)$$

then a mass balance on the gas phase gives

$$AG_g = AGx = \rho_g u_g A_g = \rho_g u_g \alpha A, \qquad (5.2)$$

where:

A = total cross-sectional area (m^2);
A_g = average cross-sectional area occupied by the gas phase (m^2); and
u_g = actual velocity of the gas phase (m/s).

Similarly for the liquid phase

$$AG_l = AG(1 - x) = \rho_l u_l A_l = \rho_l u_l (1 - \alpha)A, \qquad (5.3)$$

where:

A_l = average cross-sectional area occupied by; the liquid phase (m^2); and
u_l = actual velocity of the liquid phase (m/s).

Hence from eqns (5.2) and (5.3)

$$\alpha = \cfrac{1}{1 + \left(\cfrac{u_g}{u_l}\cfrac{1 - x}{x}\cfrac{\rho_g}{\rho_l}\right)}. \qquad (5.4)$$

For homogeneous flow the phase velocities are equal, i.e. $u_g = u_l$, and so

$$\alpha = \cfrac{1}{1 + \left(\cfrac{1 - x}{x}\cfrac{\rho_g}{\rho_l}\right)}. \qquad (5.5)$$

Fig. 5.1. Simplified picture of two-phase flow.

When ρ_l/ρ_g is large, the void fraction rises very rapidly once the quality rises even slightly above zero, as is illustrated in Fig. 5.2.

One of the main problems in two-phase flow is the calculation of the pressure gradient and homogeneous flow theory does provide answers to this problem (see § 5.3). The accuracy of the homogeneous model for pressure gradient calculations is discussed in § 5.4. First, however, single phase pressure gradients are briefly examined.

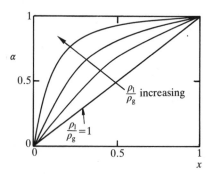

Fig. 5.2. Variation of void fraction with quality in homogeneous flow.

5.2 Single-phase pressure gradient

For flow along the tube illustrated in Fig. 5.3 the momentum equation gives, for steady flow:

pressure force + wall shear force + gravitational force

$$= \text{change in momentum flow} \quad (5.6)$$

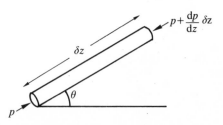

Fig. 5.3. Single-phase flow-control volume for the momentum equation.

The terms in eqn (5.6) are

$$-\frac{dp}{dz}\delta z\frac{\pi d^2}{4} - \tau\delta z\pi d - \frac{\pi d^2}{4}\delta z\rho g\sin\theta = \frac{d}{dz}(GAu)\delta z, \qquad (5.7)$$

where τ = wall shear stress (N/m^2). Because $u = G/\rho$, eqn (5.7) can now be rearranged into the convenient form

$$-\frac{dp}{dz} = \frac{4\tau}{d} + \rho g\sin\theta + G^2\frac{d}{dz}\left(\frac{1}{\rho}\right). \qquad (5.8)$$

total pressure gradient	frictional pressure gradient	gravitational pressure gradient	accelerational pressure gradient

Thus the total pressure gradient can be expressed as the sum of three components of the pressure gradient. These different components arise from distinct physical effects.

In single-phase flow the shear stress is usually expressed in terms of the friction factor

$$C_f = \frac{\tau}{\frac{1}{2}\rho u^2} = \frac{\tau}{\frac{1}{2}\frac{G^2}{\rho}}. \qquad (5.9)$$

The friction factor is a function of the Reynolds number

$$Re = \frac{\rho u d}{\mu} = \frac{Gd}{\mu}, \qquad (5.10)$$

and also, in turbulent flow, of the tube roughness ε (as in Fig. 5.4).

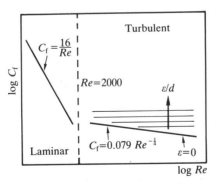

Fig. 5.4. Friction factors in single-phase flow.

5.3 Pressure gradient in homogeneous flow

In homogeneous flow, we are treating the two phases as a single fluid with a single velocity. The above analysis for single-phase flow is thus entirely valid. The only remaining questions then are:

(a) what is the appropriate homogeneous density ρ_h; and
(b) what is the appropriate homogeneous viscosity μ_h?

The homogeneous density can be found from the phase densities

$$\rho_h = \rho_g \alpha_h + \rho_l(1 - \alpha_h), \tag{5.11}$$

where α_h is the homogeneous void fraction. But this is given by eqn (5.5)

$$\alpha_h = \frac{1}{1 + \left(\dfrac{1 - x}{x}\dfrac{\rho_g}{\rho_l}\right)}.$$

So, substituting this into eqn (5.11), we obtain a simple equation for ρ_h

$$\frac{1}{\rho_h} = \frac{x}{\rho_g} + \frac{1 - x}{\rho_l}. \tag{5.12}$$

Equations for the homogeneous viscosity have been largely a matter of guesswork. The suggestions which have produced reasonable results are

$$\frac{1}{\mu_h} = \frac{x}{\mu_g} + \frac{1 - x}{\mu_l} \tag{5.13}$$

(Isbin et al. 1958),

$$\mu_h = \mu_g \frac{x\rho_h}{\rho_g} + \mu_l \frac{(1 - x)\rho_h}{\rho_l} \tag{5.14}$$

(Dukler et al. 1964), and

$$\mu_h = \mu_g \alpha_h + \mu_l(1 - \alpha_h)(1 + 2.5\alpha_h) \tag{5.15}$$

(Beattie and Whalley 1981).

These all give reasonable results for pressure gradient but the Isbin equation is generally the least good, although it is convenient to remember, being in the same form as the equation for homogeneous density (eqn (5.12)). An example of pressure gradient and void fraction calculation in homogeneous flow is given in Example 4 in Chapter 25.

5.4   When does the homogeneous model work well?

This question has to be answered separately for the three components of the total pressure gradient.

(a) *Acceleration pressure gradient*. This cannot be measured directly, but the momentum flux can be measured. The homogeneous value seems to give a reasonable prediction of the experimental results over a range of conditions, however the experimental results are rather limited. More details of accelerational pressure gradient and momentum flux are given in Chapter 6.

(b) *Gravitational pressure gradient*. This, too, cannot be measured directly, but the void fraction can be measured (see Chapter 6). It is found that the homogeneous void fraction is a good estimate of the actual void fraction if $\rho_l/\rho_g < 10$ or if $G > 2000 \text{ kg/m}^2 \text{ s}$. If these conditions are not met then the homogeneous model can under-predict the mean density by a factor of 5 to 10. It can be noted that for steam–water mixtures the condition that $\rho_l/\rho_g < 10$ corresponds approximately to $p > 120 \text{ bar}$.

(c) *Frictional pressure gradient*. This quantity also cannot be measured, but is usually obtained by subtracting the best estimates of the accelerational and gravitational terms from the total experimental pressure gradient. Fortunately it is often found that the frictional term is the dominant one. The homogeneous model, again, gives good results if $\rho_l/\rho_g < 10$ or if $G > 2000 \text{ kg/m}^2 \text{ s}$.

5.5 Two-phase multipliers

Two-phase pressure gradients are often expressed in terms of a *two-phase multiplier*. Thus

$$
\begin{array}{ccc}
\text{Two-phase pressure} \\
\text{gradient}
\end{array}
=
\begin{array}{c}
\text{single-phase} \\
\text{pressure gradient}
\end{array}
\times
\begin{array}{c}
\text{two-phase} \\
\text{multiplier.}
\end{array}
$$

Take, for example, the frictional component of the pressure gradient; this can be written as

$$
\left(-\frac{dp}{dz}\right)_F = \frac{4\tau}{d} = \frac{4}{d} C_{fh} \frac{1}{2} \frac{G^2}{\rho_h}, \tag{5.16}
$$

where C_{fh} = friction factor calculated using ρ_h and μ_h. Now eqn (5.16) can be rewritten as

$$
\left(-\frac{dp}{dz}\right)_F = \left[\frac{4}{d} C_{flo} \frac{1}{2} \frac{G^2}{\rho_l}\right]\left[\frac{C_{fh}}{C_{flo}} \frac{\rho_l}{\rho_h}\right] \tag{5.17}
$$

or

$$
\left(-\frac{dp}{dz}\right)_F = \left(-\frac{dp}{dz}\right)_{lo} \phi_{lo}^2, \tag{5.18}
$$

where here $(-dp/dz)_{lo}$ is a single phase frictional pressure gradient. The subscript $_{lo}$ means:

(a) it is a liquid single-phase flow; and

(b) it is calculated at a liquid mass flux of G = mass flux of liquid and gas in two-phase flow.

ϕ_{lo}^2 is the two-phase multiplier, i.e. the factor by which $(-dp/dz)_{lo}$ must be multiplied in order to obtain the two-phase frictional pressure gradient. Note that two-phase multipliers are usually given the symbol ϕ^2. The fact that ϕ^2 rather than ϕ is used as the symbol has no special significance: it is only an accident of the past.

In a similar way, eqn (5.16) can be rewritten in the following ways.

(a)
$$\left(-\frac{dp}{dz}\right)_F = \left[\frac{4}{d}C_{fl}\frac{G^2(1-x)^2}{\rho_l}\right]\left[\frac{C_{fh}}{C_{fl}}\frac{\rho_l}{\rho_h}\frac{1}{(1-x)^2}\right] \tag{5.19}$$

or

$$\left(-\frac{dp}{dz}\right)_F = \left(-\frac{dp}{dz}\right)_l\phi_l^2, \tag{5.20}$$

where the subscript $_l$ in $(-dp/dz)_l$ means that the flow is single-phase liquid, and that the single-phase pressure gradient is evaluated at a liquid mass flux of $G(1-x)$, which is the liquid mass flux in the two-phase flow.

(b)
$$\left(-\frac{dp}{dz}\right)_F = \left[\frac{4}{d}C_{fgo}\frac{1}{2}\frac{G^2}{\rho_g}\right]\left[\frac{C_{fh}}{C_{fgo}}\frac{\rho_g}{\rho_h}\right] \tag{5.21}$$

or

$$\left(-\frac{dp}{dz}\right)_F = \left(-\frac{dp}{dz}\right)_{go}\phi_{go}^2, \tag{5.22}$$

where the subscript $_{go}$ in $(-dp/dz)_{go}$ means that the flow is single-phase gas, and the single-phase pressure gradient is evaluated at a gas mass flux of G, which is the total mass flux in the two-phase flow.

(c)
$$\left(-\frac{dp}{dz}\right)_F = \left[\frac{4}{d}C_{fg}\frac{1}{2}\frac{(Gx)^2}{\rho_g}\right]\left[\frac{C_{fh}}{C_{fg}}\frac{\rho_g}{\rho_h}\frac{1}{x^2}\right] \tag{5.23}$$

or

$$\left(-\frac{dp}{dz}\right)_F = \left(-\frac{dp}{dz}\right)_g\phi_g^2, \tag{5.24}$$

where the subscript $_g$ in $(-dp/dz)_g$ means that the flow is single-phase gas, and the single-phase pressure gradient is evaluated at a gas mass flux of Gx, which is the gas mass flux in the two-phase flow.

The asymptotic values of the two-phase multipliers at qualities of zero and unity are given in Table 5.1.

TABLE 5.1 *Values of the two-phase multipliers at qualities of zero and one*

	ϕ_{lo}^2	ϕ_l^2	ϕ_{go}^2	ϕ_g^2
$x = 0$	1	1	$\dfrac{C_{flo}\,\rho_g}{C_{fgo}\,\rho_l}$	∞
$x = 1$	$\dfrac{C_{fgo}\,\rho_l}{C_{flo}\,\rho_g}$	∞	1	1

It can be noted that two properties make ϕ_{lo}^2 and ϕ_{go}^2 particularly convenient to use:

(*a*) the absence of any asymptote to infinity in the variation with quality; and

(*b*) the fact that $(-dp/dz)_{lo}$ and $(-dp/dz)_{go}$ are not dependent on the quality.

However, despite these desirable properties the use of ϕ_l^2 and ϕ_g^2 is still widespread.

5.6 Integration of two-phase multipliers

If we can assume that C_{fh} is approximately equal to C_{flo}, then

$$\phi_{lo}^2 = \frac{\rho_l}{\rho_h} = \frac{x\rho_l}{\rho_g} + 1 - x. \tag{5.25}$$

Since ϕ_{lo}^2 has this simple form for the homogeneous model (it is linear in quality) the integral $\int \phi_{lo}^2 \, dx$ can easily be evaluated. If the quality varies linearly with length along a heated channel then the overall frictional pressure change $(-\Delta p)_F$ can be found.

$$(-\Delta p)_F = \int \left(-\frac{dp}{dz}\right)_F dz \tag{5.26}$$

Or, using the two-phase multiplier ϕ_{lo}^2

$$(-\Delta p)_F = \left(-\frac{dp}{dz}\right)_{lo} \frac{L}{x_2 - x_1} \int_{x_1}^{x_2} \phi_{lo}^2 \, dx, \tag{5.27}$$

where:

L = length of tube (m);
x_1 = quality at tube inlet; and
x_2 = quality at tube outlet.

For homogeneous flow, it is not only the frictional component which can be integrated. Writing the equation for the total pressure gradient at a point for vertical upflow in a tube of diameter d

$$-\frac{dp}{dz} = \frac{2C_{\text{flo}}}{d}\frac{G^2}{\rho_{\text{h}}} + \rho_{\text{h}}g + G^2\frac{d}{dz}\left(\frac{1}{\rho_{\text{h}}}\right). \tag{5.28}$$

This can then be integrated to give the total pressure change Δp from an inlet quality of zero to an outlet quality of x_0. Again, it is assumed that the quality varies linearly with axial length.

$$-\Delta p = \frac{2C_{\text{flo}}LG^2}{d}\left[\frac{1}{\rho_{\text{h}}}\right] + \overline{\rho_{\text{h}}}gL + G^2\left[\frac{1}{\rho_{\text{ho}}} - \frac{1}{\rho_{\text{l}}}\right], \tag{5.29}$$

where ρ_{ho} is the homogeneous density ρ_{h} when $x = x_0$.

$$\overline{\rho_{\text{h}}} = \rho_{\text{l}}\frac{\ln(1 + X_0)}{X_0} \tag{5.30}$$

$$\left[\frac{1}{\rho_{\text{h}}}\right] = \frac{1}{\rho_{\text{l}}}\left(1 + \frac{X_0}{2}\right) \tag{5.31}$$

$$\frac{1}{\rho_{\text{ho}}} - \frac{1}{\rho_{\text{l}}} = \frac{X_0}{\rho_{\text{l}}}, \tag{5.32}$$

where, in eqns (5.30–2),

$$X_0 = x_0\left[\frac{\rho_{\text{l}} - \rho_{\text{g}}}{\rho_{\text{g}}}\right]. \tag{5.33}$$

Hence, the total pressure change can be written as

$$-\Delta p = \left(\frac{2C_{\text{flo}}L}{d}G^2\frac{1}{\rho_{\text{l}}}\right)\left(1 + \frac{X_0}{2}\right) + (\rho_{\text{l}}gL)\left(\frac{\ln(1 + X_0)}{X_0}\right) + G^2\frac{X_0}{\rho_{\text{l}}},$$
$$\tag{5.34}$$

$$\qquad\quad \text{A}\qquad\qquad \text{B}\qquad \text{C}\qquad \text{D}\qquad\qquad \text{E}$$

where the terms in eqn (5.34) represent:

A   liquid-only frictional pressure change;
B   additional term because of the two-phase flow;
C   liquid-only gravitational pressure change;
D   additional term because of the two-phase flow; and
E   momentum pressure change.

It can be noted that the momentum term arises solely because of the two-phase changing quality flow. Single-phase liquid flow would produce no momentum pressure change because the liquid velocity, and therefore the momentum flux, is constant.

References

BEATTIE, D. R. H. and WHALLEY, P. B. (1981). A simple two-phase frictional pressure drop calculation method. *Int. J. Multiphase Flow* **8**, 83–7.

DUKLER, A. E., WICKS, M., and CLEVELAND, R. G. (1964). Frictional pressure drops in two-phase flow. *AIChE J* **10**, 44.

ISBIN, H. S., MOEN, R. H., WICKEY, R. O., MOSHER, D. R., and LARSON, H. C. (1958). Two-phase steam–water pressure drops. *Nucl. Sci. and Eng. Conf.*, Chicago.

6

PRESSURE DROP IN TWO-PHASE FLOW—OVERALL METHODS FOR SEPARATED FLOW

6.1 Separated flow

In separated flow the phases physically flow with separate, different velocities, as illustrated schematically in Fig. 6.1. From eqn (5.4), the general equation for void fraction α is

$$\alpha = \frac{1}{1 + \left(\dfrac{u_g}{u_l}\dfrac{1-x}{x}\dfrac{\rho_g}{\rho_l}\right)}, \tag{6.1}$$

with the velocity ratio u_g/u_l often being called the slip ratio S. Thus, for homogeneous flow S is equal to unity. For separated flow S does not equal unity: it is usually greater than one, so that the gas is moving faster than the liquid phase.

6.2 Pressure gradient in separated flow

A comprehensive review of separated flow methods has been provided by Chisholm (1983). Just as in Chapter 4, where the momentum equation was applied to a single-phase flow (§ 5.2) and then to a homogeneous two-phase flow (§ 5.3), so it can be applied to the separated flow system. The control volume used is shown in Fig. 6.2.

$$\begin{array}{ccccccc} \text{pressure} & + & \text{wall shear} & + & \text{gravitational} & = & \text{change in} \\ \text{force} & & \text{force} & & \text{force} & & \text{momentum,} \end{array}$$

$$\tag{6.2}$$

and so

$$-\frac{dp}{dz}\delta z \frac{\pi d^2}{4} - \tau \delta z \pi d - \frac{\pi d^2}{4}\delta z[\alpha\rho_g + (1-\alpha)\rho_l]g \sin\theta$$

$$\tag{6.3}$$

$$\begin{array}{ccc} \text{pressure} & \text{shear} & \text{gravitational force} \\ \text{force} & \text{force} & \end{array}$$

$$= \frac{\pi d^2}{4}\frac{d}{dz}[\alpha\rho_g u_g^2 + (1-\alpha)\rho_l u_l^2]\delta z.$$

$$\text{change in momentum}$$

44

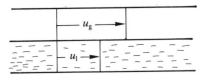

Fig. 6.1. Separated two-phase flow.

Now, using the identities

$$u_g = \frac{xG}{\alpha \rho_g}, \tag{6.4}$$

$$u_l = \frac{(1-x)G}{(1-\alpha)\rho_l}, \tag{6.5}$$

$$\rho_g u_g^2 = \frac{x^2 G^2}{\alpha \rho_g}, \tag{6.6}$$

and

$$\rho_l u_l^2 = \frac{(1-x)^2 G^2}{(1-\alpha)\rho_l}, \tag{6.7}$$

eqn (6.3) becomes

$$-\frac{dp}{dz} = \frac{4\tau}{d} + [\alpha \rho_g + (1-\alpha)\rho_l]g \sin \theta + G^2 \frac{d}{dz}\left[\frac{x^2}{\alpha \rho_g} + \frac{(1-x)^2}{(1-\alpha)\rho_l}\right]. \tag{6.8}$$

| frictional term | gravitational term | momentum or accelerational term |

It can be noted that:

(a) As with homogeneous flow, the total pressure gradient is divided into three components—frictional, gravitational, and accelerational;

(b) the gravitational and accelerational terms require a knowledge of the void fraction α;

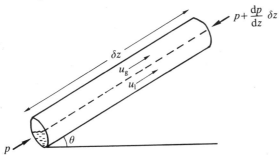

Fig. 6.2. Control volume for momentum equation analysis of separated flow.

(c) theory has, as yet, provided no help at all with the frictional term. This is often the largest of the three components.

6.3 Use of an energy equation

The steady-flow energy equation can be used as an alternative to the momentum equation for the division of the pressure gradient into components. The steady-flow energy equation is

$$\delta h + \frac{1}{2}\delta(u^2) + g \sin \theta \delta z = \delta q - \delta w, \tag{6.9}$$

enthalpy change	kinetic energy change	potential energy change

where:

q = heat transfer to the system (J/kg); and
w = external work done by the system (J/kg).

Now the change in enthalpy can be related to the change in internal energy U (J/kg)

$$\delta h = \delta U + \delta(pv), \tag{6.10}$$

where v = specific volume (m³/kg).

The change in internal energy can be obtained from the closed-system version of the first law of thermodynamics

$$\delta U = \delta q - \delta F - p\delta v, \tag{6.11}$$

where F = irreversible energy loss due to friction.

Combining eqns (6.9–11) we get

$$v\delta p + \tfrac{1}{2}\delta(u^2) + g \sin \theta \delta z + \delta F = 0. \tag{6.12}$$

Now, weighting for the mass flow rate of each phase we can write

$$v = xv_g + (1 - x)v_l \tag{6.13}$$

$$u^2 = xu_g^2 + (1 - x)u_l^2. \tag{6.14}$$

u_g and u_l can again be obtained from eqns (6.4) and (6.5). Hence eqn (6.12) can be rearranged into the form

$$-\frac{dp}{dz} = \rho_h \frac{dF}{dz} + \rho_h g \sin \theta + \tfrac{1}{2}\rho_h G^2 \frac{d}{dz}\left[\frac{x^3}{\alpha^2\rho_g^2} + \frac{(1-x)^3}{\alpha^2\rho_l^2}\right] \tag{6.15}$$

frictional term	gravitational term	momentum or accelerational term

From eqn (6.15) the following can be noted.

(a) The homogeneous density ρ_h appears because

$$\frac{1}{\rho_h} = v = xv_g + (1 - x)v_l. \tag{6.16}$$

Its appearance is accidental: it does not mean that the flow is in any way homogeneous.

(b) Three components of pressure gradient are again produced, but they are different algebraically from the momentum equation components.

(c) The total pressure gradient is of course the same but it has been divided up in a different way. This is because the momentum equation was applied to a control volume, whereas the energy equation is applied to unit mass of the flowing fluid.

The usual approach is to use the momentum equation when analysing two-phase pressure gradients. Most correlations for frictional pressure gradient and void fraction are thus based on eqn (6.8) and not on eqn (6.15).

The frictional component cannot, however, be measured directly with any accuracy. Attempts to measure the wall shear stress have been made but the results have not been very accurate. An alternative course, therefore, is to measure the void fraction (and hence to calculate the gravitational term), and to measure the momentum flux (and hence to calculate the accelerational term). The frictional term can then be found from the difference from the total measured pressure gradient.

6.4 Methods of measuring the void fraction

Experimentally, void fraction can be measured in a number of ways (see, for example, Hewitt 1978).

(a) Quick-closing valves. This is a conceptually simple and direct method (see Fig. 6.3). The relative volumes of liquid and gas trapped

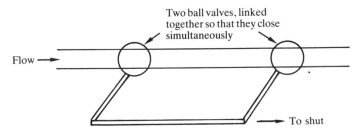

Fig. 6.3. Void fraction determination using quick-closing valves.

when the two valves are simultaneously closed enables the void fraction to be calculated.

$$\alpha = 1 - \frac{\text{volume of liquid trapped}}{\text{total volume between ball valves}} \qquad (6.17)$$

(b) *Gamma-ray densitometer.* This instrument is illustrated schematically in Fig. 6.4. For a homogeneous medium (solid metal, for example), then the intensity I of the beam varies like

$$I = I_0 e^{-\mu z}, \qquad (6.18)$$

where z = distance into the medium (m), and μ is a constant dependent upon the properties of the medium ($1/m$).

For a flow in a pipe the intensity of the beam at the detector is measured with the pipe full of liquid (I_l) and then full of gas (I_g). If, in the two-phase flow, the measured intensity is I, then what is the void fraction? We can imagine for simplicity a square pipe. In a stratified flow with the beam as shown in Fig. 6.5

$$\alpha = \frac{\ln(I/I_l)}{\ln(I_g/I_l)}, \qquad (6.19)$$

but if the beam were horizontal, then

$$\alpha = \frac{I - I_l}{I_g - I_l}. \qquad (6.20)$$

It can be seen, therefore, that the interpretation of the signals depends on the distribution of the phases. This has led to the introduction of densitometers with up to five beams which determine the void fraction distribution and so, as well as giving accurate average void fraction, also give information about the flow pattern (see Fig. 6.6). For example, for

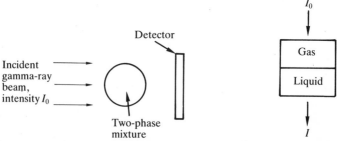

Fig. 6.4. Void fraction determination using a gamma-ray densitometer.

Fig. 6.5. Idealized square channel for gamma-ray densitometer response.

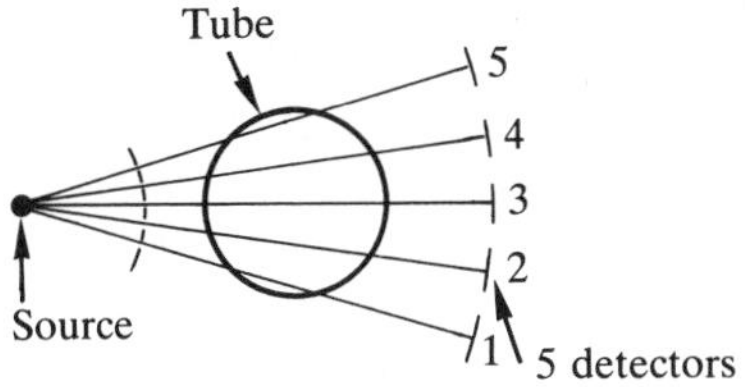

Fig. 6.6. Five-beam gamma-ray densitometer.

an annular flow detectors 1 and 5 in Fig. 6.6 would give low readings because these beams pass obliquely through the liquid film.

6.5 Correlation of void fraction

Most void-fraction correlations are actually correlations of the slip ratio S. Experimentally, it is found that slip ratio depends on (in decreasing order of importance):

(*a*) physical properties (usually expressed as ρ_l/ρ_g);
(*b*) quality x;
(*c*) mass flux G; and then
(*d*) relatively minor variables such as tube diameter, inclination of tube, length, heat flux, and flow pattern.

The dependence of slip ratio on these variables for some common correlations is summarized in Table 6.1.

TABLE 6.1 *Dependence of slip ratio S*

	S depends on		
Correlation	$\dfrac{\rho_l}{\rho_g}$	x	G
Homogeneous model, $S = 1$	no	no	no
Zivi (1964)	yes	no	no
Chisholm (1972)	yes	yes	no
Smith (1971)	yes	yes	no
CISE (Premoli *et al.* 1970)	yes	yes	yes
Zuber *et al.* (1967)	yes	yes	yes
Bryce (1977) (steam–water only)	yes	yes	yes

A number of these correlations are discussed briefly below.

(*a*) Zivi (1964) effectively assumes that total kinetic energy flow is a minimum. The kinetic energy flow in each phase is $\frac{1}{2}\rho_i u_i^2 Q_i$, where Q_i is the volume flow rate of the phase (m^3/s). Then, using the identities

$$Q_g = \frac{GxA}{\rho_g},$$

(6.21)

$$Q_l = \frac{G(1-x)A}{\rho_l},$$

(6.22)

and eqns (6.4) and (6.5) for u_g and u_l, the total kinetic energy flow

$$= \tfrac{1}{2} \rho_g \frac{G^2 x^2}{\alpha^2 \rho_g^2} \frac{GxA}{\rho_g} + \tfrac{1}{2} \rho_l \frac{G^2 (1-x)^2}{(1-\alpha)^2 \rho_l^2} \frac{G(1-x)A}{\rho_l}$$

(6.23)

$$= \frac{AG^3}{2} \left[\frac{x^3}{\alpha^2 \rho_g^2} + \frac{(1-x)^3}{(1-\alpha)^2 \rho_l^2} \right] = \frac{AG^3}{2} y,$$

(6.24)

where

$$y = \frac{x^3}{\alpha^2 \rho_g^2} + \frac{(1-x)^3}{(1-\alpha)^2 \rho_l^2}.$$

(6.25)

Now, differentiating y by α to find the minimum kinetic energy flow, we get

$$\frac{dy}{d\alpha} = \frac{-2x^3}{\alpha^3 \rho_g^2} + \frac{2(1-x)^3}{(1-\alpha)^3 \rho_l^2} = 0.$$

(6.26)

The minimum therefore occurs when

$$\frac{\alpha}{1-\alpha} = \frac{x}{1-x} \left(\frac{\rho_l}{\rho_g} \right)^{\frac{2}{3}}.$$

(6.27)

Comparison with eqn (6.1) gives

$$S = \frac{u_g}{u_l} = \left(\frac{\rho_l}{\rho_g} \right)^{\frac{1}{3}}.$$

(6.28)

The slip ratio is therefore assumed to depend only on the phase density ratio.

(b) Chisholm (1972) produced a particularly simple correlation, which is

$$S = \left[1 - x \left(1 - \frac{\rho_l}{\rho_g} \right) \right]^{\frac{1}{2}}.$$

(6.29)

Both the Zivi and the Chisholm expressions give $S \to 1$ as $\rho_l/\rho_g \to 1$, that is, as the critical point is approached (where the phase densities are equal) the flow becomes homogeneous in character. The Chisholm correlation also gives

$$S \to 1 \qquad \text{as} \quad x \to 0$$

and

$$S \to (\rho_l/\rho_g)^{\frac{1}{2}} \qquad \text{as} \quad x \to 1.$$

This latter limit for S is actually the condition that the momentum flow is a minimum.

The performance of the correlations has been assessed in comparison with a wide range of data. The order of the accuracy in calculating the mean density $\bar{\rho} = (1 - \alpha)\rho_l + \alpha\rho_g$ was: Bryce (for steam–water only); CISE; Chisholm; Smith; Zuber; Zivi; and homogeneous. It can be noted that the Chisholm correlation provides a very simple, reasonably accurate result. The most accurate generally applicable correlation is the CISE correlation: the details of this correlation are given in Appendix A. The standard deviation of the mean density calculated by the CISE correlation is approximately 40 %, although it is a little less for steam–water. An illustration of the use of the CISE correlation is given in Example 4 in Chapter 25.

6.6 Methods of measuring the momentum flux

The momentum flux has been measured as illustrated in Fig. 6.7. From the momentum equation, because all the momentum in the flow direction is destroyed, the momentum flux (flow per unit area) M (kg/m s^2) is given by

$$M = \frac{F}{A}. \qquad (6.30)$$

A large series of experiments has not been performed, but there have

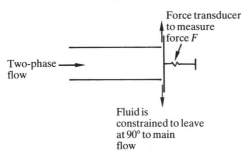

Fig. 6.7. Schematic diagram of apparatus for momentum flux measurement.

been experiments for a reasonable range of variables, for example:

fluid: steam–water
$1 \, \text{bar} < p < 10 \, \text{bar}$
$12 \, \text{mm} < d < 25 \, \text{mm}$.

One way to plot the results is as the dimensionless number $M\rho_g/G^2$ against the quality x. Then:

(a) for single-phase gas flow ($x = 1$), $M\rho_g/G^2 = 1$;
(b) for single-phase liquid flow ($x = 0$), $M\rho_g/G^2 = \rho_g/\rho_1$
(c) the homogeneous model gives

$$M = \frac{G^2}{\rho_h} = G^2 \left[\frac{x}{\rho_g} + \frac{1-x}{\rho_1} \right] \tag{6.31}$$

so

$$\frac{M\rho_g}{G^2} = x + (1-x)\frac{\rho_g}{\rho_1}. \tag{6.32}$$

Hence, for homogeneous flow, $M\rho_g/G^2$ is linear with quality. The separated model gives

$$M = G^2 \left[\frac{x^2}{\alpha\rho_g} + \frac{(1-x)^2}{(1-\alpha)\rho_1} \right], \tag{6.33}$$

and so

$$\frac{M\rho_g}{G^2} = \frac{x^2}{\alpha} + \frac{(1-x)^2}{1-\alpha}\frac{\rho_g}{\rho_1}. \tag{6.34}$$

Eliminating the void fraction from eqn (6.34) by using eqn (6.1), it can be shown that

$$\frac{M\rho_g}{G^2} = x^2 + \frac{x(1-x)}{S} + \frac{\rho_g}{\rho_1}[1 + x(S-2) + x^2(1-S)]. \tag{6.35}$$

The shape of typical curves is illustrated in Fig. 6.8.

6.7 Results for momentum flux

Typical results (from Andeen and Griffith 1968) are shown in Fig. 6.9, plotted as suggested in § 6.6. The data points cluster around the homogeneous line. The separated flow line, however, corresponds to a slip ratio known to work well for the calculation of void fraction under the experimental conditions. Why, then, does the separated flow model work badly for momentum flux, but the homogeneous model works well? The flow cannot be both homogeneous and separated at the same time.

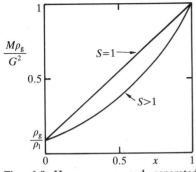

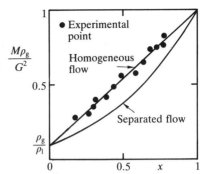

Fig. 6.8. Homogeneous and separated flow: dimensionless momentum flux variation with quality.

Fig. 6.9. Experimental results for momentum flux compared with the homogeneous and separated flow results.

The answer to this problem is that the good prediction of the homogeneous model is probably a fortuitous accident. It has not been taken into account that the real flow is not steady with time: different phase distributions at different times allow the mass flux to vary (this is most marked in plug flow). Also, there is a velocity profile across the tube (see Fig. 6.10). Larger values of G contribute excessively to the momentum flux, which is an area and time average of G^2, hence we are concerned with $\overline{G^2}$ which is larger than $(\bar{G})^2$.

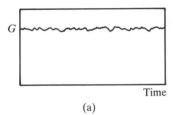

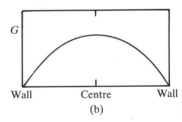

Fig. 6.10. Variation of mass flux with time and position.

Thus, although separated flow may give a good result locally and instantaneously, this averaging problem means that the overall result is too low. Fortuitously, homogeneous flow seems to give a result for the momentum flux of about the correct value.

6.8 Frictional pressure gradient

The frictional pressure gradient can be correlated by various values of ϕ_l^2, ϕ_{lo}^2, ϕ_g^2, and ϕ_{go}^2 as for homogeneous flow, the actual values of the ϕ^2 multipliers usually being determined experimentally. A useful parameter is the Martinelli parameter X^2, which is given by

$$X^2 = \left(\frac{dp}{dz}\right)_1 \bigg/ \left(\frac{dp}{dz}\right)_g. \tag{6.36}$$

Note that, as with the various ϕ^2 definitions, X^2 is normally used rather than X. $(dp/dz)_1$ and $(dp/dz)_g$ are the single-phase liquid and gas pressure gradients (N/m^3) calculated using the actual phase flow rates for the two-phase flow.

It has been found convenient to write

$$\phi_1^2 = 1 + \frac{C}{X} + \frac{1}{X^2}, \tag{6.37}$$

where C is a parameter first introduced by Chisholm. Equation (6.37) is equivalent to

$$\phi_g^2 = 1 + CX + X^2. \tag{6.38}$$

Substitution into either of these equations gives

$$\left(-\frac{dp}{dz}\right)_F = \left(-\frac{dp}{dz}\right)_1 + C\left[\left(-\frac{dp}{dz}\right)_1\left(-\frac{dp}{dz}\right)_g\right]^{\frac{1}{2}} + \left(-\frac{dp}{dz}\right)_g \tag{6.39}$$

two-phase   liquid term   'two-phase' term   gas term
frictional
pressure
gradient

The magnitude of C can be used to assess how important the specifically two-phase effects are in determining the two-phase frictional pressure drop.

First, we can work out the value of C for homogeneous flow. To simplify matters, consider turbulent flow in a rough tube where the friction factor C_f is a constant. Hence

$$\left(-\frac{dp}{dz}\right)_F = \frac{4C_f}{d}\frac{G^2}{2\rho_h}, \tag{6.40}$$

$$\left(-\frac{dp}{dz}\right)_1 = \frac{4C_f}{d}\frac{(1-x)^2G^2}{2\rho_1}, \tag{6.41}$$

and

$$\left(-\frac{dp}{dz}\right)_g = \frac{4C_f}{d}\frac{x^2G^2}{\rho_g}. \tag{6.42}$$

Therefore

$$\phi_1^2 = \frac{(-dp/dz)_F}{(-dp/dz)_1} = \frac{\rho_1}{\rho_h(1-x)^2}. \tag{6.43}$$

Now,

$$\frac{1}{\rho_h} = \frac{x}{\rho_g} + \frac{1-x}{\rho_l}, \tag{6.44}$$

and therefore

$$\phi_l^2 = \frac{\rho_l}{(1-x)^2} \left[\frac{x}{\rho_g} + \frac{1-x}{\rho_l} \right]. \tag{6.45}$$

From eqns (6.41) and (6.42)

$$X^2 = \frac{(-dp/dz)_l}{(-dp/dz)_g} = \frac{(1-x)^2}{x^2} \frac{\rho_g}{\rho_l}. \tag{6.46}$$

Substituting eqns (6.45) and (6.46) into eqn (6.47), the equation is satisfied when

$$C = \left(\frac{\rho_l}{\rho_g} \right)^{\frac{1}{2}} + \left(\frac{\rho_g}{\rho_l} \right)^{\frac{1}{2}}. \tag{6.47}$$

This is therefore the value of C for homogeneous flow, and represents an upper limit on C which occurs when the interaction effects between the phases are very large. Note that for atmospheric pressure air–water flow $C = 28.6$.

Another extreme example is a separated flow with no interaction at all between the phases: imagine each phase flowing in a separate tube of diameter d_l and d_g. Their combined cross-sectional area must be equal to the area of the tube of diameter d. The single-phase pressure gradients in each of the tubes are equal and are also equal to the frictional two-phase pressure gradient. Hence again for the fully rough turbulent flow case where the friction factor, C_f is a constant

$$\left(-\frac{dp}{dz} \right)_F = \frac{2C_f G^2 x^2}{d_g \alpha^2 \rho_g} = \frac{2C_f G^2 (1-x)^2}{d_l (1-\alpha)^2 \rho_l} \tag{6.48}$$

also from the definition of the void fraction

$$\frac{(1-\alpha)^2}{\alpha^2} = \frac{d_l^4}{d_g^4} \tag{6.49}$$

and so

$$\left(\frac{d_l}{d_g} \right)^5 = \frac{\rho_g (1-x)^2}{\rho_l x^2} \tag{6.50}$$

but

$$\left(-\frac{dp}{dz} \right)_F = \frac{2C_f G^2 x^2}{d \rho_g} \phi_g^2 = \frac{2C_f G^2 (1-x)^2}{d \rho_l} \phi_l^2 \tag{6.51}$$

therefore

$$\frac{\phi_g^2}{\phi_l^2} = \frac{\rho_g(1-x)^2}{\rho_l x^2} = \left(\frac{d_l}{d_g}\right)^5.$$ (6.52)

The total cross-sectional areas must be the same as that of a tube of diameter d

$$d_l^2 + d_g^2 = d^2$$ (6.53)

or, using eqn (6.50),

$$\frac{1}{\phi_l^{\frac{5}{2}}} + \frac{1}{\phi_g^{\frac{5}{2}}} = \frac{d^2}{d_l^2}\frac{1}{\phi_l^{\frac{5}{2}}}.$$ (6.54)

Now, from

$$\left(-\frac{dp}{dz}\right)_F = \frac{2C_f}{d_l}\frac{G^2(1-x)^2}{(1-\alpha)^2\rho_l} = \frac{2C_f}{d}\frac{G^2(1-x)^2}{\rho_l}\phi_l^2,$$ (6.55)

and using

$$\left(\frac{1}{1-\alpha}\right)^2 = \frac{d^4}{d_l^4},$$ (6.56)

then

$$\frac{d^2}{d_l^2}\frac{1}{\phi_l^{\frac{5}{2}}} = 1.$$ (6.57)

So, substituting eqn (6.57) into (6.54) we get

$$\frac{1}{\phi_l^{\frac{5}{2}}} + \frac{1}{\phi_g^{\frac{5}{2}}} = 1.$$ (6.58)

From the definitions of ϕ_l^2 and ϕ_g^2

$$\left(-\frac{dp}{dz}\right)_l^{\frac{2}{5}} + \left(-\frac{dp}{dz}\right)_g^{\frac{2}{5}} = \left(-\frac{dp}{dz}\right)_F^{\frac{2}{5}},$$ (6.59)

or

$$\left(-\frac{dp}{dz}\right)_l\left[1 + \frac{1}{X^{\frac{4}{5}}}\right]^{\frac{5}{2}} = \left(-\frac{dp}{dz}\right)_F.$$ (6.60)

Now, comparing eqn (6.60) with a slightly rearranged version of eqn (6.37)

$$\left(-\frac{dp}{dz}\right)_l\left[1 + \frac{C}{X} + \frac{1}{X^2}\right] = \left(-\frac{dp}{dz}\right)_F.$$ (6.61)

When $X = 0$ eqns (6.60) and (6.61) are identical.
When $X = \infty$ eqns (6.60) and (6.62) are identical.

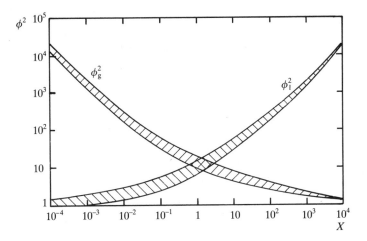

Fig. 6.11. Lockhart–Martinelli frictional pressure drop correlation.

When $X = 1$ the equations are again the same if

$$2^{\frac{5}{2}} = 2 + C \tag{6.62}$$

or

$$C = 3.66.$$

The analysis can be repeated for laminar flow, in which case $C = 2$. Note that for either turbulent or laminar flow, the values of C are much less than for homogeneous flow.

One well-known empirical correlation is that of Lockhart and Martinelli (1949) (see Fig. 6.11). The shaded regions contain four lines depending on whether each phase, flowing alone, would be laminar or turbulent. It has been shown by Chisholm (1967) that the Lockhart–Martinelli curves are well represented by the values of C in Table 6.2.

TABLE 6.2   *C values for different types of flow*

liquid	gas	C value
turbulent	turbulent	20
laminar	turbulent	12
turbulent	laminar	10
laminar	laminar	5

It can be noted that the turbulent–turbulent value for C is much bigger than the no-interaction value of 3.66. Hence, in real separated flow there is considerable interaction between the phases.

6.9 Practical correlations for frictional pressure gradient

It is not the intention here to write out the detailed form of the correlations, but to give some comments on them.

(a) The correlation of Lockhart and Martinelli (1949) (see § 6.8) is simple but not very accurate.

(b) Martinelli and Nelson (1948) give values of ϕ_{lo}^2, but only for steam–water flow. Their method is not very accurate.

(c) Thom (1964) gives values of ϕ_{lo}^2 but only for steam–water flow above 17 bar. The method gives reasonable results.

(d) Baroczy (1966) produced a peculiar graphical correlation for ϕ_{lo}^2, which is applicable to any fluid. Interpolation between the meandering lines on the graphs is difficult but it does give reasonable results.

(e) Chisholm (1973) gives an analytical equation for ϕ_{lo}^2 which is applicable to any fluid; however it is not particularly accurate.

(f) Friedel (1979) gives a complicated empirical equation for ϕ_{lo}^2 which is applicable to any fluid, except when $\mu_l/\mu_g > 1000$. It is the best generally available and generally applicable correlation. It is given in detail in Appendix B (see also Example 4 in Chapter 25). Even this optimized correlation gives a root-mean-square error of the order of 40 %. It seems unlikely that substantial improvements in this accuracy will be made until modelling methods based on individual flow patterns and their characteristics are used.

6.10 Effect of tube roughness

In single-phase flow when the fluid is turbulent, increasing the tube roughness increases the frictional pressure gradient.

There is some evidence (see Chisholm 1978) that in two-phase flow the effect of roughness is not as great as in single-phase flow. There is a possible explanation for this (see Fig. 6.12): to the gas in a two-phase flow, the apparent roughness seen is that of the liquid film, not of the

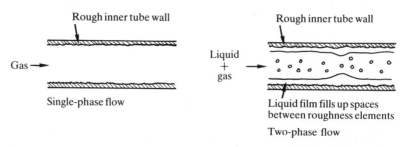

Fig. 6.12. Effect of tube-wall roughness on single-phase and two-phase pressure drop.

wall. It is therefore reasonable that as long as the tube walls are wet, the tube-wall roughness has relatively little effect.

References

ANDEEN, G. B. and GRIFFITH, P. (1968). Momentum flux in two-phase flow. *J. Heat Transfer* **90**, 211–22.

BAROCZY, C. J. (1966). A systematic correlation for two-phase pressure drop. *Chem. Eng. Prog. Symp. Ser.* **62** (no. 64), 232–49.

BRYCE, W. M. (1977). A new flow-dependent slip correlation which gives hyperbolic steam–water mixture equations. *AEEW-R1099*.

CHISHOLM, D. (1967). A theoretical basis for the Lockhart–Martinelli correlation for two-phase flow. *Int. J. Heat Mass Transfer* **10**, 1767–78.

CHISHOLM, D. (1972). An equation for velocity ratio in two-phase flow. *NEL Report 535*.

CHISHOLM, D. (1973). Pressure gradients due to friction during the flow of evaporating two-phase mixtures in smooth tubes and channels. *Int. J. Heat Mass Transfer* **16**, 347–58.

CHISHOLM, D. (1978). Influence of pipe surface roughness on friction pressure gradient during two-phase flow. *J. Mech. Eng. Sci.* **20**, 353–4.

CHISHOLM, D. (1983). *Two-phase flow in pipelines and heat exchangers*. George Godwin, London.

FRIEDEL, L. (1979). Improved friction pressure drop correlations for horizontal and vertical two-phase flow, *European Two-Phase Flow Group Meeting, Ispra, Italy* (quoted by Hewitt, 1982).

HEWITT, G. F. (1978). *Measurement of two-phase flow parameters*. Academic Press, London.

HEWITT, G. F. (1982). Pressure Drop and Void Fraction. *Handbook of Multiphase Systems* (ed. G. Hetsroni) McGraw-Hill, New York.

LOCKHART, R. W. and MARTINELLI, R. C. (1949). Proposed correlation of data for isothermal two-phase two-component flow in pipes. *Chem. Eng. Prog.* **45**, 39–48.

MARTINELLI, R. C. and NELSON, D. B. (1948). Prediction of pressure drop during forced circulation boiling of water. *Trans. ASME* **70**, 695–702.

PREMOLI, A., FRANCESCO, D., and PRINA, A. (1970). An empirical correlation for evaluating two-phase mixture density under adiabatic conditions. *European Two-Phase Flow Group Meeting, Milan* (quoted by Hewitt, 1982).

SMITH, S. L. (1971). Void fraction in two-phase flow: a correlation based upon an equal velocity head model. *Heat and Fluid Flow* **1** (No. 1), 22–39.

THOM, J. R. S. (1964). Prediction of pressure drop during forced circulation boiling of water. *Int. J. Heat Mass Transfer* **7**, 709–24.

ZIVI, S. M. (1964). Estimation of steady-state steam void fraction by means of the principle of minimum entropy production. *Trans. ASME (J. Heat Transfer)* **86**, 247–52.

ZUBER, N., STAUB, F. W., BIJWAARD, G., and KROEGER, P. G. (1967). Steady-state and transient void fraction in two-phase flow systems. *General Electric Report GEAP-5417*.

7

DRIFT FLUX MODEL

7.1 Introduction

The drift flux model is a type of separated flow model which looks particularly at the relative motion of the phases, and was developed by G. B. Wallis. The most complete reference available to the method is undoubtedly Wallis (1969). It is most applicable to flows where there is a well-defined velocity in the gas phase, for example in bubbly flow and plug flow. It is not particularly relevant to a flow like annular flow which has two characteristic velocities in one phase—the liquid film velocity and the liquid drop velocity. Nevertheless it has been used for annular flows, though with no particular success.

7.2 Development of the model

The model can be confusing due to the large number of interrelated variables used; it should be noted that u_l and u_g are the actual phase velocities (m/s), V_l and V_g are the superficial phase velocities (m/s) (the superficial velocity is the velocity the phase would have if it were flowing alone in the channel), and u_s is the slip velocity (m/s) $= u_g - u_l$. The sign convention for all the velocities is that an upward velocity is positive and a downward velocity is negative.

Now, from the above definition of the slip velocity

$$u_s = \frac{V_g}{\alpha} - \frac{V_l}{1 - \alpha},\qquad(7.1)$$

and so

$$u_s\alpha(1 - \alpha) = V_g(1 - \alpha) - V_l\alpha.\qquad(7.2)$$

The drift flux j_{gl} (m/s) of the gas relative to the liquid is defined by

$$j_{gl} = u_s\alpha(1 - \alpha)\qquad(7.3)$$

and the drift velocities, of the gas relative to the mean fluid u_{gj} (m/s), and of the liquid relative to the mean fluid u_{lj} (m/s) are defined by

$$u_{gj} = u_g - j\qquad(7.4)$$

and

$$u_{lj} = u_l - j,\qquad(7.5)$$

60

where j (m/s) is

$$j = V_1 + V_g. \tag{7.6}$$

It can be seen that the drift velocities are the difference between the actual velocity and the average velocity j. The drift flux is the volumetric flux of a component relative to the surface moving at the average velocity, thus the drift flux of the gas is

$$j_{gl} = \alpha u_{gj} = \alpha(u_g - j) \tag{7.7}$$

Substitution from eqn (7.6) and replacement of αu_g by V_g gives

$$j_{gl} = (1 - \alpha)V_g - \alpha V_1. \tag{7.8}$$

Similarly, the drift flux of the liquid can be defined as

$$j_{lg} = (1 - \alpha)u_{1j} = (1 - \alpha)(u_1 - j) \tag{7.9}$$

or

$$j_{lg} = (1 - \alpha)u_1 - (1 - \alpha)(V_g + V_1). \tag{7.10}$$

Replacement of $(1 - \alpha)u_1$ by V_1 then gives

$$j_{lg} = -(1 - \alpha)V_g + \alpha V_1 = -j_{gl}. \tag{7.11}$$

Thus the two drift fluxes j_{gl} and j_{lg} are equal and opposite; commonly only j_{gl} is used and it is usually called simply the drift flux. In upwards flow with upward velocities being defined as positive, j_{gl} is a positive quantity. Homogeneous flow, which is a flow with zero slip velocity, thus corresponds to a flow with $j_{gl} = 0$.

7.3 The physical importance of the drift flux

For steady-state one-dimensional flow (see Fig. 7.1) we can write force balances for each phase. For the liquid in the absence of wall shear stress

$$\frac{dp}{dz} + \rho_1 g - \frac{F}{1 - \alpha} = 0, \tag{7.12}$$

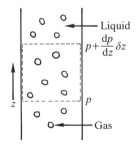

Fig. 7.1. Control volume for force balance.

and for the gas

$$\frac{dp}{dz} + \rho_g g + \frac{F}{\alpha} = 0. \tag{7.13}$$

Here F is the drag force per unit volume of mixture exerted by one phase on the other. So, eliminating dp/dz from eqns (7.12) and (7.13) we get

$$F = \alpha(1 - \alpha)(\rho_l - \rho_g)g \tag{7.14}$$

In the absence of wall shear, F is a function only of the void fraction, physical properties and the relative motion. Using eqn (7.3), F can be written as

$$F = \frac{j_{gl}}{u_s}(\rho_l - \rho_g)g. \tag{7.15}$$

Therefore both the drift flux j_{gl} and the slip velocity u_s are functions only of α and of the physical properties of the system.

7.4 Example—bubbly flow

We have already seen in Chapter 3 that, for bubbly flow, one equation for u_s is

$$u_s = u_b(1 - \alpha), \tag{7.16}$$

where u_b is the rising velocity of a single bubble (m/s). Therefore, using eqn (7.3),

$$j_{gl} = u_b \alpha(1 - \alpha)^2 \tag{7.17}$$

This function (eqn (7.17)) has been plotted in Fig. 7.2 as a function of the void fraction α, for one particular value of u_b. The graph summarizes the idea that j_{gl} is a function only of α and of the system physical properties

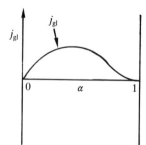

Fig. 7.2. Drift flux (eqn (7.17)) plotted as a function of void fraction.

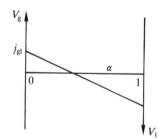

Fig. 7.3. Drift flux (eqn (7.8)) plotted as a function of void fraction.

which determine u_b. It can be noted that since

$$j_{gl} = u_s \alpha (1 - \alpha), \qquad (7.3)$$

and u_s is always a finite non-zero quantity, then

$$j_{gl} \rightarrow 0 \quad \text{as} \quad \alpha \rightarrow 0,$$
$$\text{and} \quad \alpha \rightarrow 1.$$

We also have a different type of equation for the drift flux

$$j_{gl} = (1 - \alpha)V_g - \alpha V_l, \qquad (7.8)$$

which is a consequence of a volume continuity equation for the (assumed) incompressible phases. In eqn (7.8), when:

$$\alpha = 0 \quad \text{then} \quad j_{gl} = V_g,$$
$$\alpha = 1 \quad \text{then} \quad j_{gl} = -V_l,$$

and j_{gl} is linear in α.

Hence another graphical representation of the drift flux j_{gl} is shown in Fig. 7.3. The graph summarizes the continuity equation for the system. Combining Figs. 7.2 and 7.3, the two lines for j_{gl} intersect; this point gives the actual value of α. The composite graph is shown as Fig. 7.4: essentially it is a graphical solution of eqns (7.17) and (7.8). In Fig. 7.4 the system represented is that of co-current upflows because both superficial velocities (V_g and V_l) are positive. It can be seen from Fig. 7.4 that, for this case, increasing the gas velocity V_g leads to an increase in the void fraction, and increasing the liquid velocity V_l leads to a decrease in the void fraction. Other cases can also be considered.

(a) The liquid flows up and the gas flows down (see Fig. 7.5). Here there is no solution because the situation is not physically possible.

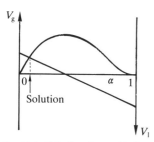

Fig. 7.4. Solution for void fraction for co-current upflow.

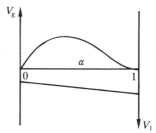

Fig. 7.5. No solution for void fraction for counter-current flow (gas flowing down, liquid flowing up).

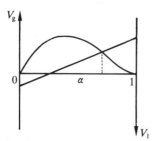

Fig. 7.6. Solution for void fraction for co-current downflow.

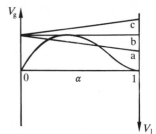

Fig. 7.7. Solution for void fraction for counter-current flow (gas flowing up, liquid flowing down).

(*b*) The liquid flows down and the gas flows down (see Fig. 7.6). A solution is obtained.

(*c*) The liquid flows down and the gas flows up (see Fig. 7.7). Here the situation is more complicated. For line c, corresponding to a large downward liquid velocity, there is no solution. For line a, corresponding to a small downward liquid velocity, there are two solutions; normally the one at lower void fraction is found. Line b must clearly represent a limit to the counter-current flow. This limit is known as flooding. (A more general description of flooding will be given in Chapter 11.) The possible areas of operation and the flooding locus are shown in Fig. 7.8. The possible values of the superficial velocities at flooding are

$$V_g = u_b 2\alpha^2(1 - \alpha), \qquad (7.18)$$

and

$$V_l = -u_b(1 - 2\alpha)(1 - \alpha)^2. \qquad (7.19)$$

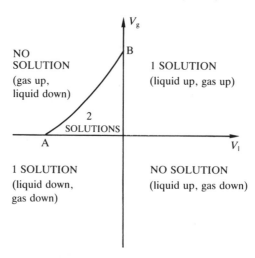

Fig. 7.8. Regions of possible operation and the flooding curve.

Equations (7.18) and (7.19) satisfy the continuity equation (eqn 7.8) and produce a line tangential to eqn (7.17). Referring to Fig. 7.8:

at A $\quad V_g = 0$, so $\alpha = 0$ and $V_l = -u_b$; and
at B $\quad V_l = 0$, so $\alpha = \frac{1}{2}$ and $V_g = u_b/4$.

Point A corresponds to a dilute suspension of bubbles being held stationary by a downflow of liquid. Point B would not normally be attainable because the bubbles tend to coalesce before a void fraction of 0.5 is reached (see Chapter 3).

7.5 Example—plug flow

From Chapter 3, the plug velocity u_p is given (if the tube is not very small) by

$$u_p = 0.35(gd)^{\frac{1}{2}}, \tag{7.20}$$

and the drift velocity of the gas relative to the mean fluid u_{gj} is

$$u_{gj} = u_p \tag{7.21}$$

If, however, we now recall from Chapter 3 that the plug actually responds to the centre-line velocity, then we should have

$$u_{gj} = 0.2j + u_p \tag{7.22}$$

if the centre-line velocity is 20 % greater than the mean velocity. The corresponding drift flux, using eqn (7.7), is

$$j_{gl} = \alpha(0.2j + u_p). \tag{7.23}$$

Substituting from eqn (7.6) for j, and from eqn (7.8) for j_{gl}, then

$$(1 - 1.2\alpha)V_g - 1.2\alpha V_l = \alpha u_p. \tag{7.24}$$

For the particular case of zero liquid flow ($V_l = 0$)

$$V_g = \frac{\alpha u_p}{1 - 1.2\alpha}. \tag{7.25}$$

This equation was previously obtained in Chapter 3. Rearranging eqn (7.24) to obtain a general equation for the void fraction in the presence of a liquid flow as well as a gas flow, we get

$$\alpha = \frac{V_g}{1.2(V_g + V_l) + u_p}. \tag{7.26}$$

7.6 More complete corrections due to profile effects

In the previous section an approximate correction method for the effects of the velocity profile was given. Of course, in bubbly flow the void fraction varies across the tube just as the velocity does.

The void fraction could be written as

$$\alpha = \frac{V_g}{u_g}. \tag{7.27}$$

Now, since

$$u_g = j + u_{gj}, \tag{7.28}$$

eqn (7.27) can be written as

$$\alpha = \frac{V_g}{j + u_{gj}} = \frac{V_g/j}{\left(1 + \dfrac{u_{gj}}{j}\right)}. \tag{7.29}$$

Equation (7.29) can be used as an implicit method of calculating the void fraction. If, for example,

$$j_{gl} = u_b \alpha (1 - \alpha)^2, \tag{7.17}$$

then

$$u_{gj} = j_{gl}/\alpha = u_b(1 - \alpha)^2 \tag{7.30}$$

and

$$\alpha = \frac{V_g/j}{1 + u_b(1 - \alpha)^2 j}. \tag{7.31}$$

However, the effect of velocity profiles and void fraction profiles across the tube has been neglected. In particular,

$$\overline{(\alpha j)} \neq (\bar{\alpha})(\bar{j}). \tag{7.32}$$

Zuber and Findlay (1965) suggested introducing a distribution parameter C_0, which is equal to

$$C_0 = \frac{\overline{(\alpha j)}}{(\bar{\alpha})(\bar{j})} \tag{7.33}$$

Zuber and Findlay suggested that for vertical upflow

$$\alpha = \frac{V_g/j}{C_0 + u_{gj}/j}, \tag{7.34}$$

where $C_0 = 1.13$. For bubbly flow they suggested that

$$u_{gj} = 1.4\left[\frac{\sigma g(\rho_1 - \rho_g)}{\rho_1^2}\right]^{\frac{1}{4}}. \tag{7.35}$$

It can be noted that this value of the drift velocity is dependent only upon the physical properties and not upon the void fraction. This would give $j_{gl} = \alpha u_{gj}$, and this does not obey the condition that $j_{gl} \to 0$ as $\alpha \to 1$. Nevertheless, it gives quite good results in the low-void-fraction region ($\alpha < 0.3$), where bubbly flow actually occurs. Equation (7.35) for u_{gj} is equivalent (for bubbly flow) to putting the rise velocity u_b of a single bubble as

$$u_b = 1.4 \left[\frac{\sigma g(\rho_l - \rho_g)}{\rho_l^2} \right]^{\frac{1}{4}}. \tag{7.36}$$

For low-pressure air–water flow this expression has a value of 0.23 m/s, which is very near the experimental result for equivalent diameters in the range 1 mm to 10 mm (see Fig. 3.4). For steam–water flow the value of the bubble rise velocity changes only slowly with pressure, at least in the range 1 bar to 100 bar (see Table 7.1).

TABLE 7.1 *Bubble rise velocity for various pressures of steam–water flow*

p (bar)	u_b (m/s)
1	0.22
3	0.21
10	0.20
30	0.19
100	0.16
221.2 (critical)	0

Near the critical point u_b falls rapidly because $\sigma \to 0$ and $\rho_g \to \rho_l$.

References

WALLIS, G. B. (1969). *One dimensional two-phase flow.* McGraw-Hill, New York.

ZUBER, N. and FINDLAY, J. A. (1965). Average volumetric concentration in two-phase flow systems. *J. Heat Transfer* **87**, 453–68.

8

CRITICAL TWO-PHASE FLOW

8.1 Single-phase flow

As an introduction, critical (or choked) single-phase flow is first reviewed briefly. Then, in subsequent sections two-phase flow is treated in an analogous manner. Critical two-phase flow is here only considered for a one-component flow, that is for a liquid and its vapour (for example a steam–water flow). From Chapter 5 eqn (5.8) gives

$$-\frac{dp}{dz} = \frac{4\tau}{d} + \rho g \sin \theta + G^2 \frac{d}{dz}\left(\frac{1}{\rho}\right). \tag{8.1}$$

This equation was derived using the momentum equation. Now put

$$\frac{d}{dz}\left(\frac{1}{\rho}\right) = -\frac{1}{\rho^2}\frac{dp}{dz} = -\frac{1}{\rho^2}\frac{dp}{dz}\left(\frac{\partial \rho}{\partial p}\right)_s,$$

where $(\partial \rho/\partial p)_s$ is evaluated at constant entropy. The use of the partial differential at constant entropy can only be justified in frictionless flow. The flow here is not frictionless, although the effects of friction are small (particularly when the flow is choked or nearly choked). Rearranging eqn (8.1) gives

$$-\frac{dp}{dz}\left[1 + \frac{G^2}{\rho^2}\left(\frac{\partial \rho}{\partial p}\right)_s\right] = \frac{4\tau}{d} + \rho g \sin \theta \tag{8.2}$$

or

$$-\frac{dp}{dz} = \frac{\left(\dfrac{4\tau}{d} + \rho g \sin \theta\right)}{\left[1 - \dfrac{G^2}{\rho^2}\left(\dfrac{\partial \rho}{\partial p}\right)_s\right]} \tag{8.3}$$

Now, using specific volume $v\ (=1/\rho)$ instead of the density, we find

$$-\frac{dp}{dz} = \frac{\left(\dfrac{4\tau}{d} + \rho g \sin \theta\right)}{\left[1 + G^2\left(\dfrac{\partial v}{\partial p}\right)_s\right]}. \tag{8.4}$$

From either eqns (8.3) or (8.4), it can be seen that the pressure gradient becomes infinite when the denominator is equal to zero. The flow is then choked. Hence choking, or critical flow, occurs when

$$1 - \frac{G_c^2}{\rho^2}\left(\frac{\partial \rho}{\partial p}\right)_s = 0, \tag{8.5}$$

where G_c = critical mass flux (kg/m^2 s). Now, $G/\rho = u$ (the fluid velocity), so choking occurs when

$$u^2 = \left(\frac{\partial p}{\partial \rho}\right)_s. \tag{8.6}$$

But $(\partial p/\partial \rho)_s$ is of course related to the velocity a (m/s) of sound, where

$$a^2 = \left(\frac{\partial p}{\partial \rho}\right)_s. \tag{8.7}$$

Further, for an ideal gas, where γ is the ratio of specific heat at constant pressure C_p to the specific heat at constant volume C_v,

$$\left(\frac{\partial p}{\partial \rho}\right)_s = \frac{\gamma p}{\rho} = \gamma RT, \tag{8.8}$$

where T = absolute temperature (K), and R = gas constant (J/kg K).

8.2   Two-phase flows

The simplest model of two-phase flow is homogeneous flow, but even in this case there is a choice to be made when critical flow is being studied. The critical mass flux can be calculated making various assumptions about what happens to the vapour in the mixture as it expands and tends to cool. The extreme assumptions which can be made are as follows.

(*a*) The quality does not change during the flow. Here the liquid is assumed not to vaporize, and so the liquid and vapour are no longer in thermodynamic equilibrium. This is homogeneous frozen flow.

(*b*) The quality is assumed to change, the liquid vaporizes and the liquid and vapour are always in equilibrium. This is homogeneous equilibrium flow.

8.3   Homogeneous frozen flow

From eqn (8.4), it can be seen that choking occurs when

$$1 + G_c^2\left(\frac{\partial v}{\partial p}\right)_s = 0, \tag{8.9}$$

where now

$$v = xv_g + (1-x)v_l. \tag{8.10}$$

If the quality x is constant then

$$\left(\frac{\partial v}{\partial p}\right)_s = x\left(\frac{\partial v_g}{\partial p}\right)_s + (1-x)\left(\frac{\partial v_l}{\partial p}\right)_s. \tag{8.11}$$

If the liquid is assumed to be incompressible, then

$$\left(\frac{\partial v_l}{\partial p}\right)_s = 0, \tag{8.12}$$

and the choking condition is

$$1 + G_c^2 x\left(\frac{\partial v_g}{\partial p}\right)_s = 0. \tag{8.13}$$

If we can also assume that the vapour behaves as an ideal gas, then

$$\left(\frac{\partial v_g}{\partial p}\right)_s = -\frac{v_g}{\gamma p} \tag{8.14}$$

and hence, at choking the mass flux is given by

$$G_c^2 = \frac{\gamma p}{x v_g}. \tag{8.15}$$

However, this equation is not very helpful as we usually do not know p and v_g at the choking plane. The situation is usually as shown in Fig. 8.1, when p_0 and v_{g0} are known. Again, single-phase results are taken; the steady-flow energy equation gives

$$\frac{T_0}{T} = 1 + \left(\frac{\gamma - 1}{2}\right)M^2, \tag{8.16}$$

where $M =$ the Mach number (u/a) and $M = 1$ at choking. Therefore, at choking,

$$\frac{T_0}{T} = \frac{1 + \gamma}{2}. \tag{8.17}$$

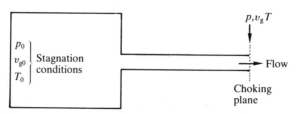

Fig. 8.1. Schematic diagram of apparatus for critical mass flux.

Therefore, assuming isentropic flow in the vapour phase,

$$\frac{p_0}{p} = \left(\frac{1+\gamma}{2}\right)^{\frac{\gamma}{\gamma-1}} \tag{8.18}$$

and

$$\frac{v_{g0}}{v_g} = \left(\frac{2}{1+\gamma}\right)^{\frac{1}{\gamma-1}} \tag{8.19}$$

Using eqns (8.18) and (8.19) for p and v_g, from eqn (8.15) we get

$$G_c^2 = \frac{p_0\gamma}{xv_{g0}}\left(\frac{2}{1+\gamma}\right)^{\frac{\gamma+1}{\gamma-1}}. \tag{8.20}$$

The equation for homogeneous frozen flow more usually quoted is that given by Starkman $et\ al.$ (1964) and Henry and Fauske (1971), which can be written in the form

$$G_c^2 = \frac{xv_{g0}p_0\gamma}{v^2}\left(\frac{2}{\gamma+1}\right), \tag{8.21}$$

where

$$v = (1-x)v_{10} + xv_{g0}\left(\frac{2}{\gamma+1}\right)^{\frac{1}{1-\gamma}}. \tag{8.22}$$

Now, when

$$xv_{g0} \gg (1-x)v_{10}, \tag{8.23}$$

eqns (8.20) and (8.21) become the same. Equation (8.21) represents an attempt to take into account, at least in part, the effects of the liquid and so remove the unfortunate tendency of eqn (8.20) to give unreasonably large answers at very low qualities. Equation (8.21) is most easily proved by starting from the steady-flow energy equation, considering the enthalpy change of the vapour phase, and assuming that the critical pressure ratio is the same as in single-phase flow. This latter assumption is proved by Henry and Fauske (1971) for the case where inequality (8.23) is true.

8.4 Homogeneous equilibrium flow

Once again, at the choking plane, eqn (8.9) applies.

$$1 + G_c^2\left(\frac{\partial v}{\partial p}\right)_s = 0, \tag{8.9}$$

and again

$$v = xv_g + (1-x)v_1. \tag{8.10}$$

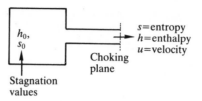

Fig. 8.2. Control volume for steady-flow energy equation.

However, the quality is no longer constant, and so substituting eqn (8.10) into eqn (8.9) gives

$$1 + G_c^2\left[\frac{dx}{dp}(v_g - v_l) + x\frac{dv_g}{dp} + (1 - x)\frac{dv_l}{dp}\right] = 0, \qquad (8.24)$$

but this is not very helpful as x is not known at the choking plane.

An alternative approach to the study of homogeneous equilibrium flow is to use the steady-flow energy equation as a starting point (see Fig. 8.2). The stagnation enthalpy h_0 is given by

$$h_0 = h + \frac{u^2}{2}, \qquad (8.25)$$

or in terms of the specific volume

$$h_0 = h + \frac{G^2 v^2}{2}. \qquad (8.26)$$

Therefore

$$G_c = \frac{1}{v}[2(h_0 - h)]^{\frac{1}{2}}. \qquad (8.27)$$

Assuming once again that the flow is isentropic,

$$s_0 = (1 - x)s_l + xs_g, \qquad (8.28)$$

and so

$$x = \frac{(s_0 - s_l)}{(s_g - s_l)}, \qquad (8.29)$$

where s_g and s_l are the saturation gas and liquid entropies (J/kg K). Then

$$h = (1 - x)h_l + xh_g, \qquad (8.30)$$

where h_g and h_l are the saturation gas and liquid enthalpies (J/kg), and

$$v = (1 - x)v_l + xv_g, \qquad (8.31)$$

where v_g and v_l are the saturation gas and liquid specific volumes (m³/kg).

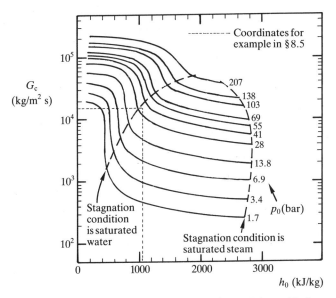

Fig. 8.3. Homogeneous equilibrium flow: chart for determining critical mass flux for steam–water flow.

The procedure for calculating the critical mass flux is:

(a) calculate h_0 and s_0 from property tables;

(b) guess a pressure, p at the choking plane;

(c) look up s_1 and s_g from tables at the pressure p and then calculate the quality x from eqn (8.29);

(d) calculate h and v from eqns (8.30) and (8.31) after first obtaining the values of h_1, h_g, v_1, and v_g from property tables;

(e) calculate the mass flux from eqn (8.27);

(f) vary the pressure p and repeat steps (c), (d) and (e); until

(g) the maximum value of the mass flux is found. This is the critical mass flux.

This procedure is long and tedious. For the special case of steam–water flow the results have been expressed in the form of a chart where the critical mass flux G_c is expressed as a function of the stagnation pressure p_0 and the stagnation quality x_0. This chart from Moody (1965) is shown in Figure 8.3.

8.5 Example (see Fig. 8.4)

Steam and water at a stagnation pressure p_0 of 34 bar and a quality of 0.01 is contained in the vessel as shown. Find the critical mass flux in the outlet pipe which discharges to the atmosphere.

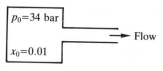

Fig. 8.4. Example.

From steam tables

$$v_{l0} = 1.231 \times 10^{-3} \, \text{m}^3/\text{kg}$$

and

$$v_{g0} = 0.0587 \, \text{m}^3/\text{kg},$$

and, for steam,

$$\gamma = 1.3.$$

Using the frozen-flow model the quality is equal to x_0, and so from eqn (8.20), putting $p_0 = 34 \times 10^5 \, \text{N/m}^2$, $G_c = 50\,700 \, \text{kg/m}^2 \, \text{s}$.

From eqn (8.22) $v = 2.154 \times 10^{-3} \, \text{m}^3/\text{kg}$, and then, from eqn (8.21), $G_c = 22\,000 \, \text{kg/m}^2 \, \text{s}$.

Note that the two versions of the frozen-flow model give, for this example, significantly different results. This is because the inequality (8.23) is not satisfied because the quality is very low.

Using the homogeneous equilibrium model the results of the trial-and-error solution are shown in Fig. 8.5. The critical mass flux is 15 100 kg/m^2 s, and the value of the pressure p_c at the choking plane is 30 bar. The corresponding quality at the choking plane is 2.8 %. It can be noted

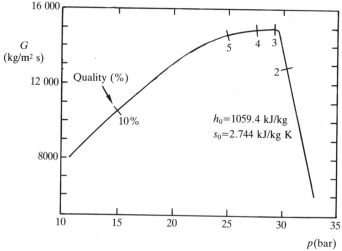

Fig. 8.5. Result of hand calculation for critical mass flux using homogeneous equilibrium flow model.

that the frozen-flow models give the critical pressure ratio to be

$$\left(\frac{2}{\gamma+1}\right)^{\frac{\gamma}{\gamma-1}} = 0.55.$$

For the above example the critical pressure ratio according to the equilibrium-flow model is $30/34 = 0.88$. Generally, the equilibrium model gives the best prediction of critical pressure ratio, and the frozen model (eqn (8.21)) gives the best result for the critical mass flux. For the case considered here the experimental results were $G_c = 24\,000\ \text{kg/m}^2\ \text{s}$ and a critical pressure ratio equal to 0.83.

As the quality increases, all the models give similar results for the mass flux and similar results for the pressure ratio. For example, if the above example were repeated with dry saturated steam in the vessel (so that $x_0 = 1.00$), then the results shown in Table 8.1 would be obtained.

TABLE 8.1 *Values of G_c and critical pressure ratio for dry, saturated steam*

	G_c $(\text{kg/m}^2\ \text{s})$	p_c/p_0
Frozen-flow model	5077 (from both eqn (8.20) and eqn (8.21))	0.55
Equilibrium-flow model	4810	0.57

In addition, the quality predicted at the choking plane by the equilibrium model was 0.945. A further example of the calculation of critical flow rates in two-phase flow is given in Example 5 in Chaper 25.

8.6 Other models of the flow

Other assumptions about the nature of the flow are clearly possible (see Levy 1965). In particular, we can ask the following questions.

(a) Is the flow actually homogeneous? Often this is not too bad an assumption as the velocities are very high and so the flow is very well dispersed. This tends to give the phases equal velocities.

(b) Is the flow frozen or equilibrium? It seems that real flows are neither. There is some change in quality (so it is not frozen), but not as much in equilibrium flow.

8.7 Non-homogeneous flow

The homogeneous model momentum flux is G^2/p or G^2v, where

$$v = v_1(1 - x) + v_g x.$$

Alternatives are the separated flow equations.

(a) From the momentum equation the momentum flux is given by

$$\text{momentum flux} = G^2 \left[\frac{x^2 v_g}{\alpha} + \frac{(1-x)^2 v_1}{1-\alpha} \right]. \tag{8.32}$$

The void fraction α is then usually calculated using a slip ratio S which minimizes the momentum flux.

$$\alpha = \frac{1}{\left[1 + S\left(\frac{1-x}{x} \right) \frac{\rho_g}{\rho_1} \right]} \tag{8.33}$$

$$S = \left(\frac{\rho_1}{\rho_g} \right)^{\frac{1}{2}}. \tag{8.34}$$

(b) From the energy equation

$$\text{momentum flux} = G^2 \left[\frac{x^3 v_g}{\alpha^2} + \frac{(1-x)^3 v_1}{(1-\alpha)^2} \right]. \tag{8.35}$$

Then the void fraction α is usually calculated (see Moody, 1965) using a slip ratio which minimizes the kinetic energy flux

$$S = \left(\frac{\rho_1}{\rho_g} \right)^{\frac{1}{3}}. \tag{8.36}$$

8.8 Experimental results

Figures 8.6 and 8.7 show some typical experimental results for the situation illustrated in Fig. 8.1. The fluid was steam–water and the stagnation pressure was 34 bar. Note how the various methods predict critical mass fluxes which are similar if the quality is large and also, at high qualities, how the critical pressure ratio becomes similar. The Henry–Fauske (1971) theory combines non-equilibrium flow (treated in a rather empirical manner) and non-homogeneous flow: eqns (8.33) and (8.34) are used.

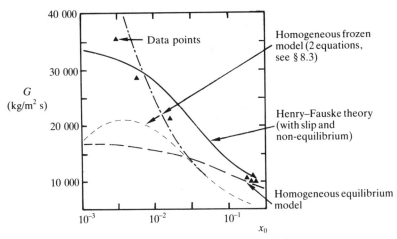

Fig. 8.6. Typical experimental results for critical mass flux for steam–water flow at $p_0 = 34$ bar.

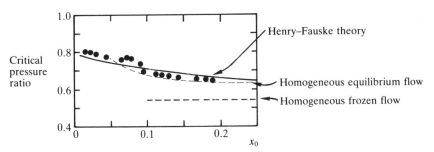

Fig. 8.7. Typical experimental results for critical pressure ratio for steam–water flow at $p_0 = 34$ bar.

References

HENRY, R. E. and FAUSKE, H. K. (1971). The two-phase critical heat flow of one-component mixtures in nozzles, orifices and short tubes. *J. Heat Transfer* **93**, 179–87.

LEVY, S. (1965). Prediction of two-phase critical flow rate. *J. Heat Transfer* **87**, 53–8.

MOODY, F. I. (1965). Maximum flow rate of a single component two-phase mixture. *J. Heat Transfer* **87**, 134–42.

STARKMAN, E. S., SCHROCK, V. E., NEUSEN, K. F., and MANEELY, D. F. (1964). Expansion of a very low quality two-phase fluid through a convergent–divergent nozzle. *J. Basic Engineering* **86**, 247–56.

9

DETAILED PHENOMENA IN ANNULAR FLOW: WAVE BEHAVIOUR, DEPOSITION, AND ENTRAINMENT

9.1 Introduction

Chapter 4 looked at annular flow in an overall manner, but few of the details of the flow were considered. Here, the detailed phenomena of wave behaviour and droplet behaviour are reviewed. Droplet behaviour, in particular the rate at which droplets are formed and destroyed (entrainment and deposition), enables a detailed physical model of annular flow to be constructed (see Chapter 10).

9.2 Disturbance waves

One of the distinctive features of annular flow is the presence of disturbance waves. They are quite different from waves on a falling film because the ratio of the wave amplitude to the mean film thickness is large (see Fig. 9.1). Typical values of the film thickness and wave parameters for an air–water flow are:

minimum film thickness $= 96\ \mu$m;
mean film thickness $= 315\ \mu$m;
maximum film thickness $= 1810\ \mu$m;
wave velocity $= 3.25$ m/s;
wave frequency $= 14.5$ Hz;
actual velocity of air $= 21.3$ m/s; and
average actual velocity of water film $= 1.2$ m/s.

Disturbance waves occur in almost all practical applications. As a rough rule, it can be said that they occur when the liquid film Reynolds number Re_{lf} is greater than 200 (Cousins et al. 1965). In this case

$$Re_{\text{lf}} = \frac{G_{\text{lf}}d}{\mu_{\text{l}}}, \qquad (9.1)$$

where:

$d =$ tube diameter (m);
$\mu_{\text{l}} =$ liquid viscosity (N s/m^2); and
$G_{\text{lf}} =$ superficial liquid film mass flux mass $=$ flow rate in the liquid film/tube cross-sectional area.

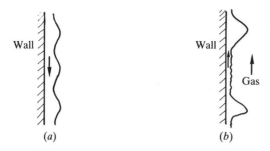

Fig. 9.1. Waves on liquid films: (a) falling-film wave; and (b) annular-flow disturbance wave.

For air–water flows in transparent tubes the waves appear as milky bands about 50 mm long covering the whole of the tube circumference. Closer examination shows that for small diameter tubes ($d < 50$ mm) the waves do indeed have a well-defined circumferential identity. However, for tubes of larger diameter ($d > 100$ mm) there is no circumferential identity at all, and the waves are localized to relatively small areas. Figure 9.2 is a schematic diagram of the view normal to the tube axis of an air–water flow in a 125 mm diameter tube.

When the flow is viewed from inside the tube, the picture is very different. Such views have been obtained in two different ways (see Whalley *et al.* 1979).

(a) *Side illumination with the camera focussed on the plane of illumination* (*see Fig. 9.3*). This has the advantage that it gives a perspective view of the flow, and the method can be used with a long tube. The disadvantage is the large amount of light required, and the depth of field is rather small. Because the depth of field is small, a droplet or wave can only be followed for a short distance.

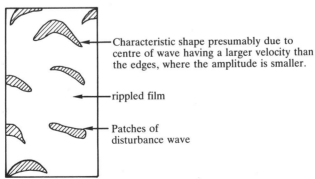

Fig. 9.2. Waves observed by viewing a tube of 125 mm diameter from the side (Azzopardi and Gibbons 1983).

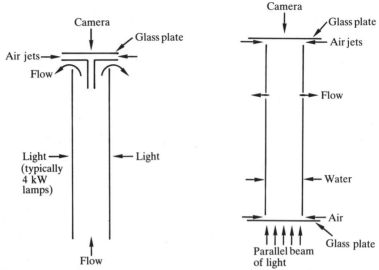

Fig. 9.3. Axial view photography: apparatus using side lighting.

Fig. 9.4. Axial view photography: apparatus using parallel-light illumination.

(b) *Axial illumination using a parallel beam of light with the camera focussed on infinity (see Fig. 9.4).* This method has the advantage that only a small amount of light is necessary (10 mW He/Ne laser) and a large depth of field (up to 0.5 m) can be obtained. The disadvantages are that it gives no sense of perspective and cannot be used with a tube longer than about 2 m.

The results of both techniques, when used with high-speed cine-photography (up to 4000 frames per second), show that each disturbance wave is extremely complex. Within the wave there are many sub-peaks and troughs. The most striking feature seen on the cine films is the presence of filaments of liquid which eventually shatter, leaving drops behind (see Fig. 9.5). This process appears to be the main source of entrainment of drops in the flow. The flow is extremely complex and

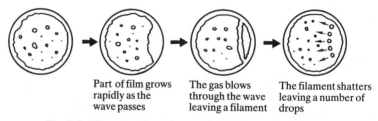

Part of film grows rapidly as the wave passes

The gas blows through the wave leaving a filament

The filament shatters leaving a number of drops

Fig. 9.5. Film rupture as observed by axial-view photography.

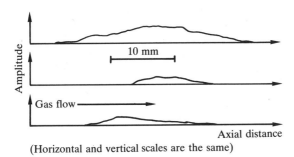

Fig. 9.6. Wave profiles measured by Hewitt and Nicholls (1969).

highly turbulent in nature. There seems, at the moment, little hope of understanding in detail the processes at work: certainly the conditions are far removed from most encountered in conventional fluid mechanics (see Lighthill 1978).

The wave shape profile has been measured, though inevitably with some smoothing of the very disturbed and rough interface seen in the cine films. By measuring the intensity of the excitation light emitted by a fluorescent dye dissolved in the liquid, giving a signal which is proportional to the film thickness at a point, the wave profile has been determined. Typical results for air–water flow are shown in Fig. 9.6. The smoothed profile of the wave is rather flat; the wave is quite long compared with its amplitude. Commonly, it is found that the ratio of amplitude of the wave to the mean liquid film thickness is about 5:1. The amplitude of the waves is thus large compared with the film on which they are travelling. It is because the waves have, relatively, such large amplitudes that it is difficult to make much theoretical progress in understanding their behaviour.

9.3 Wave frequency and velocity

The wave frequency in a long vertical tube is initially high and then drops rapidly (see Fig. 9.7). At the point of liquid inlet the frequency seems to be approximately independent of the liquid flow rate. The drop in frequency is caused by wave coalescence. Coalescence occurs frequently near the liquid inlet. Subsequently, the frequency tends to an asymptotic value. The coalescence has been studied by measuring wave positions at various times (see the results illustrated in Fig. 9.8). The events illustrated by this figure are:

(a) Two waves coalesce to form one wave with a larger velocity than either of the initial waves;

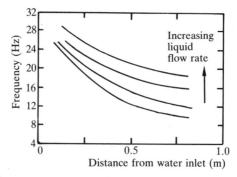

Fig. 9.7. Wave frequency as a function of distance from the inlet (from Hewitt and Lovegrove 1969).

(b) Two waves approach each other and then appear to repel each other; and

(c) Two waves merge to form one high-velocity wave which subsequently slows down.

Two events in particular were not observed.

(a) Waves were never seen to break up so that one wave formed two waves. Sometimes, however, waves do decay until they become indistinguishable from the many ripples on the film.

(b) Waves were never seen to pass through each other—instead they coalesce. This observation is significant because it has been suggested that waves are solitons—solitary wave solutions of the Korteweg–de Vries equation (see Lighthill 1978). Solitons, however, have the property of passing through each other virtually unchanged.

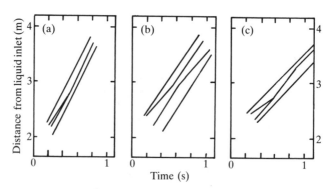

Fig. 9.8. Wave trajectories (from Nedderman and Shearer 1963).

The velocity of the waves increases with both liquid flow rate and gas flow rate, although the gas velocity has a larger effect than the liquid flow rate. Typical velocities are usually in the range 2 to 4 m/s for air–water. The velocity changes little except in the entrance region, where the frequency is changing rapidly. One largely empirical equation for wave velocity u_w which works well was given by Pearce (1979)

$$u_w = \frac{u_s + u_g(\rho_g/\rho_l)^{\frac{1}{2}}}{1 + (\rho_g/\rho_l)^{\frac{1}{2}}},$$ (9.2)

where: u_g is the actual gas velocity (m/s); u_s is the surface velocity of the film (m/s); and ρ_g and ρ_l are the gas and liquid densities (kg/m³). Taking the typical values given in § 9.2, and putting the surface velocity of the film equal to twice the average film velocity, this equation gives $u_w = 3.04$ m/s, whereas the experimental result is 3.25 m/s.

The wave separation, sometimes misleadingly called the wavelength, is given by the wave velocity divided by the wave frequency. For the typical data given in § 9.2, the separation is 3.25/14.5 = 0.22 m. The wave spacing, in general, is asymptotic to a minimum value as the liquid flow rate increases. The asymptote decreases roughly linearly with increasing gas flow rate (see Figs. 9.9 and 9.10). This is only the average spacing between the waves as the velocity and frequency measurements themselves are averages. Individual spacings between successive waves determined, for example from cine films, can be presented as a probability distribution as in Fig. 9.11. The positions A, B, C, and D in Fig. 9.11 represent peaks (or possible peaks) in the distribution. Note that A, B, C, and D are approximately equally spaced. Also, for this tube, the circumference πd is 80 mm, so the positions of A, B, C, and D correspond to $n\pi d$ for n equal to one, two, three, and four.

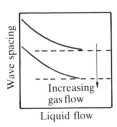

Fig. 9.9. Wave spacing as a function of liquid flow rate.

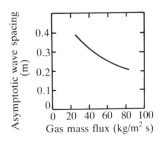

Fig. 9.10. Asymptotic wave spacing as a function of gas flow rate.

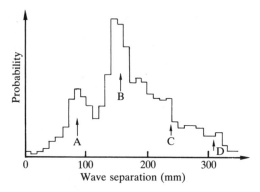

Fig. 9.11. Probability density function of wave separation (from Hall Taylor and Nedderman 1968).

9.4 Artificial waves

In order to aid the study of wave behaviour 'artificial' waves have been created and studied. This has been done for annular flow by setting up a situation with the liquid flow just below the value at which waves are first formed. Then extra liquid is injected quickly from a syringe into the liquid film. The main results from artificial wave studies are summarized below (see Azzopardi and Whalley 1980).

(*a*) If the liquid flow is significantly below that required for wave formation, then the injected liquid forms a wave which slowly dies away.

(*b*) In all respects (amplitude, velocity, and appearance) the artificial waves behave like real, naturally occurring waves.

(*c*) The volume of liquid required to form each wave is approximately constant and about equal to (for a tube diameter of 0.032 m) 3 ml. If 6 ml of liquid is injected then two waves, one closely following the other, are formed.

(*d*) By injecting concentrated salt solution rather than pure liquid, the use of probes detecting electrical conductivity enables us to find out what happens to the salt. The experiment is illustrated in Fig. 9.12. The conductivity peak appears long after the wave has passed the probe at B, therefore the high conductivity liquid which was in the wave at A is no longer in it at B.

This latter result is quite conventional behaviour for ordinary waves: they are propagating surface disturbances, not packets of liquid moving along the surface of the liquid. However, it is still more than possible that disturbance waves in annular flow can be visualized as packets of liquid rolling along the top of the film, though with a fairly rapid interchange between the liquid in the wave and the liquid in the film.

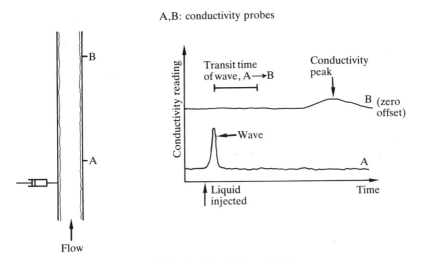

Fig. 9.12. Salt injection into single artificial waves.

9.5 Fraction of liquid entrained

It should be noted that entrainment in annular flow can mean two things: the entrainment rate of drops from the film into the gas core, or the fraction of the liquid which is entrained. The entrained liquid fraction is looked at first because it is relatively easy to measure (see Hewitt 1978a).

The amount of liquid entrained can be very significant—up to 80 % or more of the liquid in the two-phase flow can be flowing as entrained liquid drops. The usual technique of measuring the entrained liquid fraction (though actually what is measured is the liquid film flow rate), is illustrated in Fig. 9.13. The liquid film is extracted through part of the tube wall which is made of porous material. The whole of the liquid film is extracted when a small amount of the gas is taken off. Thereafter the liquid extraction rate increases only very slowly with the gas extraction rate because of the difficulty of diverting the high speed entrained drops into the sinter. As shown in Fig. 9.13, this technique works well for vertical flow, but it works less well for horizontal flow (see Butterworth 1972) or flow in a helical coil (see Whalley 1980), where the film flow is dependent upon circumferential position. In these cases partial film over part of the tube circumference is used.

The type of experiment just described gives a measure of the equilibrium fraction E of the liquid which is entrained, where

$$E = \frac{\text{liquid droplet flow rate}}{\text{total liquid flow rate}} . \qquad (9.3)$$

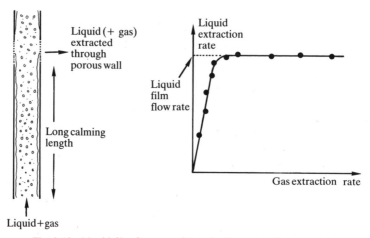

Fig. 9.13. Liquid film flow rate determination using film removal.

E varies with length and only slowly reaches a steady value, as illustrated by the data in Fig. 9.14. The following can be noted from Fig. 9.14.

(*a*) About 400 tube diameters are required for equilibrium to be finally attained. Many experiments which have been performed have not used tubes long enough for equilibrium to be reached.

(*b*) Equilibrium is never actually attained. In Fig. 9.14, dG_{le}/dz is positive. This is because the pressure of the system is continually falling in the direction of flow. This causes the gas density to decrease, the velocity to increase, and *E* therefore to increase.

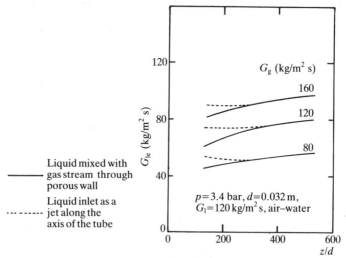

Fig. 9.14. The approach to hydrodynamic equilibrium in annular flow (from Brown *et al.* 1975).

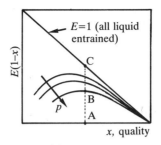

Fig. 9.15. Entrainment diagram showing variation of equilibrium entrained liquid fraction with system pressure.

The general form of results is often expressed as an *entrainment diagram*, as shown in Fig. 9.15; see for example Hewitt and Pulling (1969). This is a graph of the ratio G_{le}/G plotted against the quality. Note that

$$\frac{G_{le}}{G} = \frac{G_{le}}{G_l}\frac{G_l}{G} = E(1-x). \tag{9.4}$$

In the diagram, the diagonal line represents the total amount of liquid at any quality. Also,

(a) as noted above, E increases as the pressure falls;

(b) the length AB is proportional to the entrained liquid superficial mass flux G_{le}; and

(c) the length BC is proportional to the liquid film superficial mass flux G_{lf}.

It is evident that entrainment is a dynamic process and that the entrainment fraction is the result of the competing effects of entrainment and deposition. Thus any simple, direct correlation of entrainment fraction is almost bound to fail (or at least to work only over a very limited range of data). Paleev and Filippovich (1966) correlated $(1-E)$ with

$$\frac{\rho_{gc}}{\rho_1}\left(\frac{\mu_1 V_g}{\sigma}\right)^2,$$

where ρ_{gc} is the density of the gas core flow (kg/m^3). However data from different tube diameters and different fluids fall in different regions of the graph. Ishii and Mishima (1982) were much more successful when they put

$$E = \tanh(7.25 \times 10^{-7} V_g^{+2.5} d^{+1.25} Re_1^{0.25}), \tag{9.5}$$

$$V_g^+ = V_g\left[\frac{\rho_g^{\frac{4}{3}}}{\sigma g(\rho_1 - \rho_g)^{\frac{1}{3}}}\right]^{\frac{1}{4}}, \tag{9.6}$$

$$d^+ = d\left[\frac{g(\rho_1 - \rho_g)}{\sigma}\right]^{\frac{1}{2}}, \tag{9.7}$$

and

$$Re_1 = \frac{\rho_1 V_1 d}{\mu_1},\tag{9.8}$$

where V_1 and V_g are the superficial liquid and gas velocities (m/s). Ideally, we should look at entrainment and deposition rates and the mechanisms of these two processes. A fuller review is given by Hewitt (1978b).

9.6 Deposition of drops from the core

From photographic investigations using an axial view along the tube (see § 9.2) it is possible to observe the deposition process. It is found that the drops move in straight lines (as far as their non-axial movement is concerned) from the point where they are entrained to the point where they are deposited (see James *et al.* 1980). Figure 9.16 shows typical observed drop tracks; very occasionally a drop was seen to bounce off the liquid film rather than to deposit on to it. The initial transverse velocity of drops is of the order of $u^* = (\tau_i/\rho_g)^{\frac{1}{2}}$, the friction velocity, where τ_i = interfacial shear stress (N/m^2). However, the drops which can be seen in these axial view experiments have a diameter D_d which is greater than about 150 μm, but the mean drop size is lower than this (typically around 100 μm). The drop size decreases with increasing gas flow rate and goes through a shallow minimum as the liquid flow increases. The mean usually used is the Sauter mean diameter (sometimes called D_{d32}) which is defined by

$$D_{d32} = \frac{\sum n_i D_{di}^3}{\sum n_i D_{di}^2},\tag{9.9}$$

where there are n_i drops with diameter D_{di}. The many small drops are assumed to interact in a complex way with the gas phase turbulence. Three ranges of drop size can be distinguished.

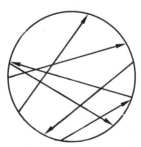

Fig. 9.16. Droplet trajectories (in the non-axial direction) observed using the axial-view parallel-light apparatus.

(a) *Diameters less than* $0.1\,\mu$m. The drops follow the gas phase turbulence very closely but only deposit after diffusing through the laminar sub-layer in the gas boundary layer. It is not clear how the laminar sub-layer is affected by the very disturbed liquid film next to it.

(b) *Diameters between* $0.1\,\mu$m and $10\,\mu$m. Drops still follow the eddies in the gas core but their inertia is now sufficient for them to penetrate the laminar sub-layer more easily.

(c) *Diameters greater than* $10\,\mu$m. All drops have sufficient inertia to penetrate the laminar sub-layer but they no longer follow the gas phase turbulence at all closely.

The literature about the deposition rate can be summarized by reference to Fig. 9.17. Here: k is the deposition mass transfer coefficient (m/s); u^* is the friction velocity $(\tau_i/\rho_g)^{\frac{1}{2}}$ (m/s); τ_i is the interfacial shear stress (N/m^2); Sc is the Schmidt number $\mu_g/\mathcal{D}\rho_g$; $\mathcal{D}$ is the diffusion coefficient due to Brownian motion (m^2/s); and τ^+ is the dimensionless relaxation time which is defined by

$$\tau^+ = \frac{D_d^2\rho_g\rho_l u^{*2}}{18\mu_g^2}. \tag{9.10}$$

The deposition rate D (kg/m^2 s), expressed as the mass flow rate deposited per unit area of tube wall, is then given by

$$D = kC, \tag{9.11}$$

where C is the concentration of liquid drops in the gas core (kg/m^3). Note that here it is implicitly assumed that the effective concentration of droplets at the tube wall is zero. This is because the tube wall is assumed

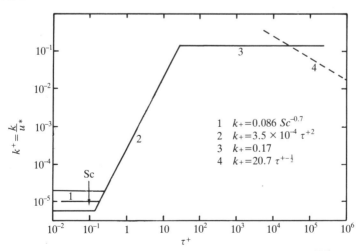

Fig. 9.17. Deposition of liquid drops (McCoy and Hanratty 1977).

to remove all the droplets hitting it. For typical air–water values:

$\rho_g = 2\,\text{kg/m}^3$,
$\rho_l = 1000\,\text{kg/m}^3$,
$\tau_i = 16.8\,\text{N/m}^2$, and
$\mu_g = 17 \times 10^{-6}\,\text{N s/m}^2$;

values of the droplet diameter at the region boundaries in Fig. 9.17 are:

$1 \rightarrow 2$ diameter $= 0.24\,\mu\text{m}$,
$2 \rightarrow 3$ diameter $= 2.6\,\mu\text{m}$, and
$3 \rightarrow 4$ diameter $= 68\,\mu\text{m}$

The value of the mass transfer coefficient k in region 3, the region containing most of the small drops in annular flow, using the above data is

$$k = 0.17u^* = 0.17(16.8/2)^{\frac{1}{2}} = 0.49\,\text{m/s}.$$

As the droplet diameter increases above $68\,\mu\text{m}$, the mass transfer coefficient decreases according to the equation in Fig. 9.17 for region 4. Measurements in annular flow show that a typical value is $0.15\,\text{m/s}$ for air–water flow. This was found by unidirectional deposition experiments of the type illustrated in Fig. 9.18. There are porous wall sections at positions A and B. The one at A removes the initial film flow and leaves

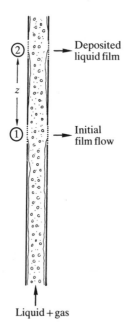

Deposited
liquid film

Initial
film flow

Fig. 9.18. Determination of deposition
by successive film removal.

Liquid + gas

a dry wall. The one at B then removes the film formed by the drops depositing over the distance z. The film at B must not be sufficiently thick for re-entrainment to occur from it. Now, neglecting the volume occupied by the liquid drops,

$$C \approx \rho_g \frac{W_{le}}{W_g}, \qquad (9.12)$$

where W_{le} is the entrained liquid flow rate (kg/s) and W_g is the gas flow rate (kg/s).

Now, consider an element of the gas core δz long, and calculate the liquid flows in and out of it.

Entrained liquid flow rate in $= W_{le}$.
Entrained liquid flow rate out $= W_{le} + (dW_{le}/dz)\, \delta z$.

Deposition flow rate out $= k\pi d\, \delta z C \approx k\pi d\, \delta z \rho_g \dfrac{W_{le}}{W_g}$. $\qquad (9.13)$

Hence, because the flow is at steady state

$$\frac{dW_{le}}{dz} = -k\pi d\rho_g \frac{W_{le}}{W_g}. \qquad (9.14)$$

Integrating and putting the boundary conditions such that $W_{le} = W_{le1}$ when $z = 0$, and $W_{le} = W_{le2}$ when $z = z$, then

$$k = \frac{W_g \ln(W_{le1}/W_{le2})}{\pi d\rho_g z} \qquad (9.15)$$

Typical experimental results for an air–water system are shown in Fig. 9.19. From this figure it can be seen that the deposition mass transfer coefficient k:

(a) asymptotes to about 0.15 m/s, and
(b) is initially higher. This is probably due to a combination of large drops depositing as a result of the initial inertia on entrainment and the establishment of an equilibrium concentration profile of drops.

Results also show that for steam–water at high pressure (≈ 70 bar), the deposition mass transfer coefficient k is much lower and is around 0.01 m/s.

It is known that the mass transfer coefficient depends on drop concentration in the gas core (see Fig. 9.20). One explanation for the drop in mass transfer coefficient with increasing droplet concentration is that the high droplet concentration reduces the turbulence level in the gas core.

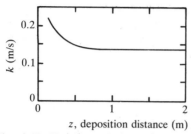

Fig. 9.19. Variation of deposition mass transfer coefficient with distance from the initial film removal point (from Cousins and Hewitt 1968).

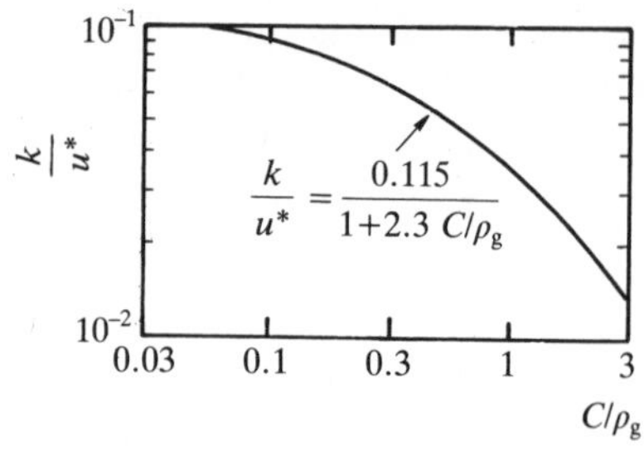

Fig. 9.20. Effect of droplet concentration on the deposition mass transfer coefficient (from Andreussi 1983).

Unidirectional deposition experiments in annular flow can be criticized because, after the film is removed at the first sinter, the character of the flow is changed because the wavy film has been removed. The turbulence level in the core will therefore not be typical of the turbulence levels in real annular flows with a wavy liquid film.

9.7 Entrainment of droplets from the liquid film

From artificial wave experiments and other experiments it is known that the presence of waves is intimately associated with the entrainment process. The actual entrainment mechanism is one where part of the wave is undercut by the gas flow. Figure 9.21 shows a side view of the process. An axial view of the process can be seen in Fig. 9.5. Unfortunately, our understanding of wave behaviour is very incomplete and it is not surprising that the entrainment process is not properly understood.

Another problem occurs because it is very difficult to measure rates of entrainment from a film in annular flow. It can be done near the entrance region in a developing flow where the liquid is introduced entirely as liquid film, and initially there are no entrained drops. The film flow rate can then be measured as a function of axial distance. As there is initially no deposition, the initial value of dG_{lf}/dz can be used to calculate the entrainment rate. These initial values of the entrainment rate are, however, not actually representative of fully developed flow. This is because the waves take some time to reach an asymptotic frequency (see Fig. 9.7) and the gas phase turbulence and velocity profile take some time to build up. Some measure of the entrainment rate can be inferred from the drops released from a single wave (most usefully an artificial wave, see § 9.4), but this is an extremely tedious experiment as necessarily only one wave can be in the apparatus at once. The only alternative is to argue that for an equilibrium flow (where G_{lf} does not change with length), the

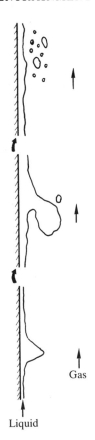

Fig. 9.21. Entrainment mechanism.

entrainment rate is equal to the deposition rate, or

$$E = D, \tag{9.16}$$

where E is the entrainment rate (kg/m^2 s) expressed as the mass flow rate of liquid entrained into the gas core per unit area of tube wall. We can now say that

$$D = kC_E, \tag{9.17}$$

where k is the deposition mass transfer coefficient (m/s) and C_E is the droplet concentration (kg/m^3) in the gas core. Hence, at equilibrium,

$$E = kC_E. \tag{9.18}$$

Furthermore, we can extrapolate and use eqn (9.18) even when the film and the gas core are not in equilibrium. Now, however, C_E is not the actual droplet concentration, but the concentration which would be in equilibrium with the liquid film. But how can this equilibrium

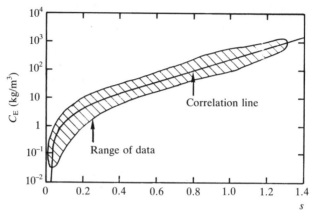

Fig. 9.22. Entrainment correlation (from Hutchinson and Whalley 1973).

concentration C_E be calculated? One idea was to argue that the liquid film was held together by surface tension forces and disrupted by the interfacial shear stress. Now, σ and τ_i can be made into a dimensionless group s if the average film thickness m is used:

$$s = \tau_i m / \sigma. \tag{9.19}$$

Then, taking data which really are equilibrium data (when the entrainment and deposition rates are equal, so that the concentration of liquid drops is the equilibrium concentration), it was found that the data could be plotted as shown in Fig. 9.22. The scatter is large, but some correlation between C_E and s seems to exist. Alternative correlations which have been tried have included (Whalley and Hewitt 1978):

(a) C_E/d against s, and
(b) $E\sigma/(\tau_i \mu_l)$ against s (E calculated from eqn 9.18),

but these showed only slight improvement over the correlation in Fig. 9.22. The group $E\sigma/(\tau_i \mu_l)$, at least, has the advantage that it is dimensionless.

References

ANDREUSSI, P. (1983). Droplet transfer in two-phase annular flow. *Int. J. Multiphase Flow* **9**, 697–713.

AZZOPARDI, B. J. and GIBBONS, D. B. (1983). Annular two-phase flow in a large diameter tube. *The Chemical Engineer* No. **398**, 19–21.

AZZOPARDI, B. J. and WHALLEY, P. B. (1980). Artificial waves in annular two-phase flow. ASME Winter Annual Meeting. In *Basic mechanisms in two-phase flow and heat transfer*. ASME.

BROWN, D. J., JENSEN, A., and WHALLEY, P. B. (1975). Non-equilibrium effects in heated and unheated annular two-phase flow. *ASME paper 75-WA/HT-7.*

BUTTERWORTH, D. (1972). Air–water annular flow in a horizontal tube. Int. Symp. on Two-Phase Systems, Haifa, Israel. In *Prog. Heat Mass Transfer* **6**, 235–51.

COUSINS, L. B. and HEWITT, G. F. (1968). Liquid phase mass transfer in annular two-phase flow: droplet deposition and liquid entrainment. *AERE-R5657.*

COUSINS, L. B., DENTON, W. H., and HEWITT, G. F. (1965). Liquid mass transfer in annular two-phase flow. Two-phase flow symposium, Exeter, UK (see also AERE-R4926).

HALL TAYLOR, N. S. and NEDDERMAN, R. M. (1968). The coalescence of disturbance waves in annular two-phase flow. *Chem. Eng. Sci.* **23**, 551–64.

HEWITT, G. F. (1978a). *Measurement of two-phase flow parameters.* Academic Press, London.

HEWITT, G. F. (1978b). Liquid mass transport in annular two-phase flow. Seminar, Int. Center Heat Mass Transfer, Dubrovnik.

HEWITT, G. F. and LOVEGROVE, P. C. (1969). Frequency and velocity measurements in annular two-phase flow. *AERE-R4304.*

HEWITT, G. F. and NICHOLLS, B. (1969). Film thickness measurement in annular two-phase flow using a fluorescence spectrometer technique. *AERE-R4506.*

HEWITT, G. F. and PULLING, D. J. (1969). Liquid entrainment in adiabatic steam–water flow. *AERE-R5374.*

HUTCHINSON, P. and WHALLEY, P. B. (1973). A possible characterization of entrainment in annular flow. *Chem. Eng. Sci.* **28**, 974–5.

ISHII, M. and MISHIMA, K. (1982). Liquid transfer and entrainment correlation for droplet-annular flow. *7th Int. Heat Transfer Conference,* Munich **5**, 307–12.

JAMES, P. W., HEWITT, G. F., and WHALLEY, P. B. (1980). Droplet motion in two-phase flow, ANS/ASME/NRC topical meeting on nuclear reactor thermal-hydraulics. *NUREG-/CP-0014* **2**, 1484–503.

LIGHTHILL, M. J. (1978). *Waves in fluids.* Cambridge University Press.

McCOY, D. D. and HANRATTY, T. J. (1977). Rate of deposition of droplets in annular two-phase flow. *Int. J. Multiphase Flow* **3**, 319–31.

NEDDERMAN, R. M. and SHEARER, C. J. (1963). The motion and frequency of large disturbance waves in annular two-phase flow of air–water mixtures. *Chem. Eng. Sci.* **18**, 661–70.

PALEEV, I. and FILIPPOVICH, B. S. (1966). Phenomena of liquid transfer in two-phase dispersed annular flow. *Int. J. Heat Mass Transfer* **9**, 1089–93.

PEARCE, D. L. (1979). Film waves in horizontal annular flow space-time correlator experiments. *CERL RD/L/N111/79.*

WHALLEY, P. B. (1980). Air–water two-phase flow in a helically coiled tube. *Int. J. Multiphase Flow* **6**, 345–56.

WHALLEY, P. B. and HEWITT, G. F. (1978). The correlation of liquid entrainment fraction and entrainment rate in annular two-phase flow. *AERE-R9187.*

WHALLEY, P. B., HEWITT, G. F., and TERRY, J. W. (1979). Photographic studies of two-phase flow using a parallel light technique. *AERE-R9389.*

10

ANNULAR FLOW MODELS

10.1 Introduction

Here we consider the physical modelling of annular flow in the absence of evaporation and condensation. So

$$\frac{dG_g}{dz} = 0,$$ (10.1)

and

$$\frac{dG_l}{dz} = 0,$$ (10.2)

where:

z = axial length (m),
G_g = superficial gas mass flux (kg/m^2 s) = Gx, and
G_l = superficial liquid mass flux (kg/m^2 s) = $G(1-x)$.

It is assumed that the physical properties are independent of z, so there is no flashing of liquid into gas (vapour). The case where there is evaporation or condensation is considered in detail in Chapter 19. Now consider an element of length δz, as shown in Fig. 10.1.

liquid film flow rate in $= \dfrac{\pi d^2}{4} G_{lf}$,

liquid film flow rate out $= \dfrac{\pi d^2}{4} \left(G_{lf} + \dfrac{dG_{lf}}{dz} \delta z \right)$,

deposition in the length $\delta z = \pi d\, \delta z D$, and
entrainment in the length $\delta z = \pi d\, \delta z E$,

where E and D are mass flow rates per unit area of tube wall. The film flow into the element equals the film flow out of the element, so

$$\frac{\pi d^2}{4} G_{lf} + \pi d\, \delta z D = \frac{\pi d^2}{4} \left(G_{lf} + \frac{dG_{lf}}{dz} \delta z \right) + \pi d\, \delta z E$$ (10.3)

or

$$\frac{dG_{lf}}{dz} = \frac{4}{d}(D - E).$$ (10.4)

A similar mass balance on the entrained liquid gives

$$\frac{dG_{le}}{dz} = \frac{4}{d}(E - D).$$ (10.5)

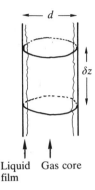

Fig. 10.1. Control volume for film-flow mass balance.

Note that

$$\frac{dG_{lf}}{dz} + \frac{dG_{le}}{dz} = 0, \qquad (10.6)$$

so that eqn (10.2) is satisfied, since

$$G_l = G_{lf} + G_{le}. \qquad (10.7)$$

10.2 Methods of use

There are two main uses of an annular flow model.

(*a*) The first main use is for integration of the equations to give G_{le} or G_{lf} as a function of z. Clearly this is an initial value problem, that is G_{le} or G_{lf} must be known as the integration is started. Note that in the integration G_{le} and G_{lf} eventually reach constant values. Then

$$\frac{dG_{le}}{dz} = 0, \qquad \frac{dG_{lf}}{dz} = 0, \quad \text{and} \quad E = D.$$

(*b*) The second main use is not to find the variation of G_{le} or G_{lf} with axial length, but to find the equilibrium flow. This occurs, as above, when the entrainment and deposition rates are equal.

10.3 Integration along the channel

If eqn (10.5)

$$\frac{dG_{le}}{dz} = \frac{4}{d}(E - D) \qquad (10.5)$$

is to be integrated, then information must be available about the entrainment rate E (kg/m^2 s) and the deposition rate D (kg/m^2 s). From Chapter 9

$$E = kC_E, \tag{10.8}$$

$$C_E = f\left(\frac{\tau_i m}{\sigma}\right), \quad \text{and} \tag{10.9}$$

$$D = kC, \tag{10.10}$$

where C is the concentration of drops in the gas core (kg/m^3), and the function in eqn (10.9) is shown in Fig. 9.22. If it is assumed that the gas and the liquid drops in the core flow as a homogeneous mixture (that is, with the same velocity u_c), then

$$C = G_{le}/u_c \tag{10.11}$$

or

$$C = \frac{G_{le}}{\left(\dfrac{G_g}{\rho_g} + \dfrac{G_{le}}{\rho_l}\right)}. \tag{10.12}$$

Note that because $\rho_l \gg \rho_g$ when G_{le} is not large compared with G_g

$$C = \frac{G_{le}\rho_g}{G_g}. \tag{10.13}$$

This latter equation was used previously in Chapter 9. For the special case of zero entrainment rate

$$\frac{dG_{le}}{dz} = -\frac{4}{d} D. \tag{10.14}$$

Putting

$$D = kC = \frac{kG_{le}\rho_g}{G_g}, \tag{10.15}$$

then

$$\frac{dG_{le}}{dz} = -\frac{4k}{d}\frac{\rho_g}{G_g} G_{le} \tag{10.16}$$

and integrating

$$G_{le} = G_{le0}e^{-4k\rho_g/dG_g z}, \tag{10.17}$$

where G_{le0} is the value of G_{le} at $z = 0$. Alternatively we can write

$$G_{le} = G_{le0}e^{-z/z_0}, \tag{10.18}$$

where z_0 is a characteristic length given by

$$z_0 = \frac{dG_g}{4k\rho_g}. \tag{10.19}$$

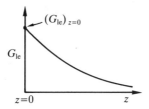

Fig. 10.2. Decay of entrained liquid mass flux in the absence of deposition.

Now the gas phase velocity u_g is given approximately by

$$u_g = G_g/\rho_g, \tag{10.20}$$

so that

$$z_0 = \frac{du_g}{4k}. \tag{10.21}$$

Equation (10.18) is an exponential decay to zero as shown in Fig. 10.2. Taking typical values:

$d = 0.025$ m,
$k = 0.15$ m/s, and
$u_g = 30$ m/s

gives

$z_0 = 1.25$ m.

This characteristic length is rather short. On the basis of deposition experiments in long tubes, longer lengths would be expected. This may be because the actual value of the mass transfer coefficient is not constant but decreases as illustrated by the data in Fig. 9.19.

10.4 Effect of entrainment

With entrainment, the integration is unfortunately much more difficult. Equation (10.5) becomes, when substitutions are made for E and D from eqns (10.8) and (10.10),

$$\frac{dG_{le}}{dz} = \frac{4k}{d}(C_E - C). \tag{10.22}$$

However C_E is a complicated function of the interfacial shear stress and average film thickness. The procedure for the calculation process is therefore as follows.

(a) Knowing G_{le}, C can be calculated from eqn (10.12).
(b) Knowing G_{le}, the interfacial shear stress τ_i and the average film

thickness m can be calculating by solving, simultaneously

    (i) the triangular relationship, and
    (ii) the interfacial roughness correlation,

which are both equations connecting G_{le}, m, and τ_i.

    (c) C_E can then be calculated from eqn (10.9) and Fig. 9.22.
    (d) Finally dG_{le}/dz can be found from eqn (10.22).

G_{le} is then obtained as a function of the axial distance z by any standard numerical integration technique, for example the Runge–Kutta method. The most complicated of these steps is step (b) involving the simultaneous solution of the triangular relationship and the interfacial roughness correlation. This step was discussed in § 4.5.

Typical results of the integration are shown in Fig. 10.3. The conditions for the experiments and the calculations in Fig. 10.3 were:

air–water, $p = 1.9$ bar,
$d = 0.032$ m,
$G_l = 158$ kg/m^2 s, and
$G_g = 79$ kg/m^2 s.

Two things are immediately apparent from Fig. 10.3.

(a) The approach to equilibrium is calculated as occurring sooner than it actually occurs. This is probably because the simple mass transfer description for the entrainment and deposition is too crude to represent the processes properly. Also, the waves take some distance (of the order of 1 m or so) to build up properly and come to an approximately constant frequency.

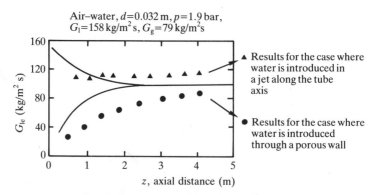

Fig. 10.3. Variation of entrained liquid mass flux with axial length: calculated and experimental results (from Hutchinson *et al.* 1974).

(b) The final equilibrium line (from 3 m onwards) seems to be a reasonable approximation to the actual final equilibrium value of entrained liquid mass flux G_{le}.

10.5 Prediction of the final equilibrium value of G_{le}

The final value of G_{le} can, of course, be predicted from the integration process described in the previous section. It is more easily obtained by putting at equilibrium

$$E = D,$$ (10.23)

and therefore

$$C = C_E.$$ (10.24)

This is now equivalent to solving three equations simultaneously for three variables:

(a) triangular relationship,
(b) interfacial roughness correlation, and
(c) entrainment correlation (eqn (10.9)).

The three variables are:

(a) interfacial shear stress τ_i,
(b) average film thickness m, and
(c) entrained liquid mass flux G_{le}.

From these variables other useful variables can be calculated, for example the pressure gradient and the void fraction.

The results of these calculations for the three variables have been compared by Hewitt and Whalley (1978) with a large collection of experimental data points. The data used included points for low-pressure air–water flow, low-pressure air–organic liquid flow and steam–water flow at pressures varying from 2.4 bar to 100 bar. The error was expressed as a percentage error, which was the root-mean-square error as a percentage of the mean value for that variables. The results are given in Table 10.1.

TABLE 10.1   *Typical errors calculated for the variables in § 10.5*

Variable	Error (%)
Entrained liquid fraction G_{le}/G_l	32
Interfacial shear stress τ_i	90
Average film thickness m	24

Two things can be noted.

(*a*) The error in the shear stress is particularly large.
(*b*) The other errors are typical of the errors encountered in calculating parameters in two-phase flow.

The equations used were:

(*a*) the triangular relationship, as given in Chapter 4;
(*b*) the interfacial roughness correlation of Wallis (1970)

$$C_{\text{fi}} = C_{\text{fg}}\left(1 + 360\frac{m}{d}\right), \quad \text{and} \tag{10.25}$$

(*c*) the entrainment correlation, eqn (10.9), and Fig. 9.22.

The large error in the calculated interfacial shear stress can be reduced by substituting the alternative interfacial roughness correlation of Hewitt and Whalley (1978)

$$C_{\text{fi}} = C_{\text{fg}}\left[1 + 24\left(\frac{\rho_{\text{l}}}{\rho_{\text{g}}}\right)^{\frac{1}{3}}\frac{m}{d}\right]. \tag{10.26}$$

The results when tested against the same set of data are given in Table 10.2.

TABLE 10.2   *Change in error produced by the use of the Hewitt–Whalley correlation*

Variable	Error (%)	Change in error
Entrained liquid fraction $G_{\text{le}}/G_{\text{l}}$	29	Minor reduction
Interfacial shear stress τ_{i}	26	Major reduction
Average film thickness m	38	Increased

Now the errors in all the variables are typical of those given by the best correlations for many two-phase flow parameters, for example frictional pressure drop.

It has been demonstrated that the more commonly needed variables of two-phase pressure gradient and void fraction can be calculated from the annular-flow models with an accuracy that is comparable to that obtained from the best correlations. Oddly enough, good results are obtained from the annular-flow modelling method whether or not the actual flow is really annular. Further details about the modelling of annular flow and the results obtained are given by Hewitt (1982).

References

HEWITT, G. F. (1982). Prediction of pressure drop in annular flow. In *Handbook of multiphase systems* (ed. G. Hetsroni). McGraw-Hill, New York.

HEWITT, G. F. and WHALLEY, P. B. (1978). The correlation of liquid entrainment fraction and entrainment rate in annular two-phase flow. *AERE-R 9187.*

HUTCHINSON, P., WHALLEY, P. B., and HEWITT, G. F. (1974). Transient flow redistribution in annular two-phase flow. *Int. J. Multiphase Flow* **1,** 383–93.

WALLIS, G. B. (1970). Annular two-phase flow, II Additional effects. *J. Basic Engineering* **92,** 73–82.

11

FLOODING IN TWO-PHASE FLOW

11.1 Introduction

The terms 'flooding' and 'flow reversal' are explained with reference to Fig. 11.1. The liquid flow rate is held constant while the gas flow rate is changed as shown. The events are then as follows.

(a) First, there is falling film flow with no gas flow. The film has ripples on its surface which grow slowly as the gas flow rate increases.

(b) As the rate of upwards flow of gas is increased the liquid film becomes progressively more disturbed.

(c) The liquid film gradually becomes very disturbed with large waves which partially block the channel. Some of the liquid is propelled by the gas flow above the liquid inlet. There are a large number of liquid drops formed in the chaotic flow. This is *flooding*.

(d) More and more of the liquid is pulled upwards by the gas, until

(e) all the liquid moves up and co-current annular flow is formed. The liquid film behaves in an oscillatory manner: it moves intermittently up and down, although the net flow is up.

(f) The gas flow is now reduced progressively.

(g) The liquid film first 'hangs' at the liquid inlet point (see Fig. 11.2). This has been termed the *hanging film*. The film then begins to flow downwards. This is the *flow reversal* point.

There is some degree of hysteresis, that is the gas velocity at flooding is greater than that at flow reversal.

11.2 Mechanism of flooding

A number of mechanisms for flooding have been proposed. The most obvious is that the rising gas exerts a shear force on the liquid film; this is analysed using the element of the film shown in Fig. 11.3. If it is assumed that the pressure variation in the gas is negligible, and therefore that the pressure variation in the liquid film is negligible, then a force balance on the film gives

$$\delta z(\tau_i + \tau) = (m - y)\rho_1 g \delta z, \tag{11.1}$$

where the symbols are defined with reference to Fig. 11.3. Note that τ and τ_i are shear stresses (N/m^2). Therefore

$$\tau = (m - y)\rho_1 g - \tau_i. \tag{11.2}$$

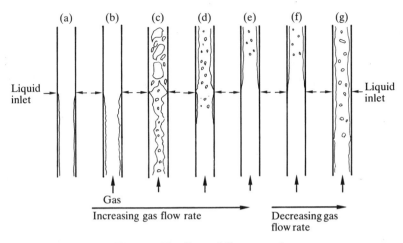

Fig. 11.1 Flooding and flow reversal.

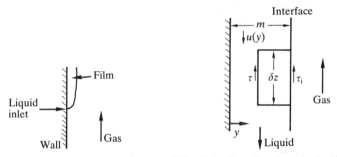

Fig. 11.2. Hanging film.

Fig. 11.3. Control volume in the liquid film.

Now, if the film is laminar,

$$\tau = \mu_1 \frac{du}{dy} \qquad (11.3)$$

and

$$\frac{du}{dy} = (m - y) \frac{\rho_1 g}{\mu_1} - \frac{\tau_i}{\mu_1}. \qquad (11.4)$$

Integrating,

$$u = \frac{\rho_1 g}{\mu_1} \left[my - \frac{y^2}{2} \right] - \frac{\tau_i y}{\mu_1} + C, \qquad (11.5)$$

where C is the constant of integration. Then $C = 0$ as $u = 0$ when $y = 0$.
A special case of eqn (11.5) occurs when the interfacial shear stress τ_i is

zero, then

$$u = \frac{\rho_1 g}{\mu_1} \left[my - \frac{y^2}{2} \right],$$ (11.6)

and volume flow per unit width in the film Q_1 (m^2/s) is given by

$$\int_0^m u \, dy = \frac{\rho_1 g}{\mu_1} \int_0^m \left(my - \frac{y^2}{2} \right) dy = \frac{\rho_1 g}{\mu_1} \frac{m^3}{6},$$ (11.7)

and thus

$$m = \left[\frac{6\mu_1 Q_1}{\rho_1 g} \right]^{\frac{1}{3}},$$ (11.8)

which is the Nusselt film thickness. Returning now to the full equation for u, eqn (11.5), the velocity profile for a fixed film thickness can be plotted as the interfacial shear stress τ_i increases (see Fig. 11.4). Point A, where $u = 0$ at $y = m$, occurs when

$$\tau_i = \frac{1}{2} \rho_1 g m.$$ (11.9)

Here the interface is stationary. Point B, where the net liquid flow is zero, is given by the solution of

$$\int_0^m u \, dy = 0,$$ (11.10)

which is

$$\tau_i = \frac{2}{3} \rho_1 g m.$$ (11.11)

It is tempting to think that this analysis (of point B, in particular) will explain flooding. However, the interfacial shear stress τ_i required to initiate flooding is well below that given by any of these formulae (see Hewitt *et al.* 1965).

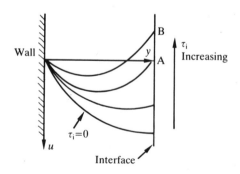

Fig. 11.4. Velocity profiles in a laminar liquid film.

The most probable mechanism involves the growth of waves on the liquid film. The evidence for this is fourfold: entry phenomena, length effects, wave injection results and visualization. This evidence is discussed in the following sections.

11.3 Entry phenomena

It is known that the detailed geometry of liquid and gas entry can affect the gas velocity at which flooding occurs. For example, in Fig. 11.5 the flooding gas velocity in apparatus A will be appreciably lower than for apparatus B. The interpretation is that apparatus A, because of the sharp-edged gas entry, has a higher level of gas-phase turbulence, thus promoting wave growth on the liquid film (see Hewitt 1982).

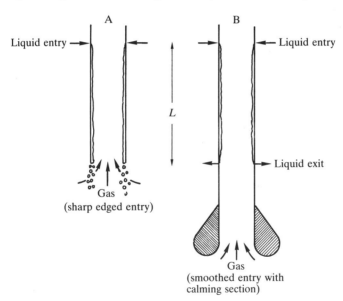

Fig. 11.5. Entry phenomena in flooding.

11.4 Length effects

There has long been disagreement about whether the tube length L has any effect on the gas velocity necessary to initiate flooding. However, it is generally found that in situations with a high turbulence level, such as apparatus A in Fig. 11.5, the tube length has no effect. Where the turbulence level is low, however, such as in apparatus B, there is a length effect, as is illustrated in Fig. 11.6. The gas velocity in flooding in long tubes is lower than that in short tubes. The interpretation is that in longer

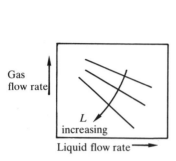

Fig. 11.6. Length effect in flooding.

Fig. 11.7. Artificial wave injection into a falling liquid film.

tubes the liquid waves have more time to build up, so the flooding occurs at lower values of the gas velocity.

11.5 Wave injection results

In § 9.4, the production of artificial single waves in co-current upflow was described. These waves behaved exactly like naturally occurring waves. Similarly, waves on the falling liquid film can be artificially induced by injecting extra liquid into the film from a syringe. Experiments have been carried out by Whalley and McQuillan (1985), who investigated the effects of injecting waves at positions A, B, and C in apparatus of the form of Fig. 11.7. The main findings were as follows.

(*a*) waves injected at A have a large effect, that is they tend to cause flooding easily. Waves injected at B, and particularly at C, have much less effect.

(*b*) Large injected waves have a greater effect than small waves.

(*c*) Large injected waves cause flooding not at the liquid exit (where normal flooding usually starts) but between the liquid entry and exit.

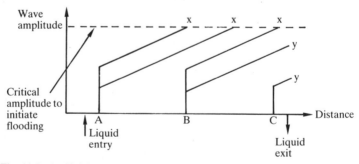

Fig. 11.8. Artificial wave injection and flooding: interpretation of the results.

It seems possible that these findings can all be explained by the wave growth mechanism (see Fig. 11.8). If there is a critical wave amplitude which is necessary to cause flooding and the injected waves grow as they fall, as illustrated in Fig. 11.8, then the three facts above are all illustrated by Fig. 11.8. Flooding is assumed to happen at the points labelled x in this figure, but not at the points labelled y.

11.6 Visualization experiments

Two series of photographic experiments have given different views of the flooding process (McQuillan *et al.* 1985).

(*a*) With an axial view down the tube, the explosive growth of a wave at flooding is seen. However, there is little information about the axial position along tube of any disturbance. An attempt to 'mark' various axial positions in the tube with coloured side light was only partially successful.

(*b*) With a conventional side view, the picture is chaotic so flooding on the inside of an annular test section was photographed (see Fig. 11.9). The central rod was actually a tube with two porous sections of wall to enable film to be formed and removed. The outer transparent shroud tube contained a venturi-like constriction which gives larger air velocities in the region where the camera is focussed. This caused flooding to occur reliably within the field of view. The results show waves on the liquid film growing and falling. Just before flooding the waves slow down and then

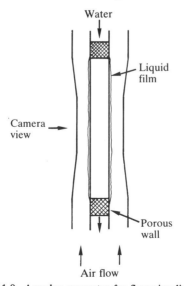

Fig. 11.9. Annulus apparatus for flow visualization.

one wave is stationary for a short time. Other waves coalesce with it as they fall into it and the wave thus grows. It is then blown upward by the gas, causing the formation of a large number of liquid droplets.

11.7 Flooding correlations

The methods of correlating flooding which are normally used are of two types.

(a) *The Wallis* (1961) *type*. The general form of this correlation is

$$V_1^{*\frac{1}{2}} + V_g^{*\frac{1}{2}} = C, \qquad (11.12)$$

where V_1^* is a dimensionless liquid superficial velocity

$$V_1^* = V_1 \rho_1^{\frac{1}{2}} / [gd(\rho_1 - \rho_g)]^{\frac{1}{2}}, \qquad (11.13)$$

V_g^* is a dimensionless gas superficial velocity given by

$$V_g^* = V_g \rho_g^{\frac{1}{2}} / [gd(\rho_1 - \rho_g)]^{\frac{1}{2}}, \qquad (11.14)$$

V_1 is the gas superficial velocity (m/s), V_g is the gas superficial velocity (m/s), and C is a constant which has a value of about 0.8.

Equation (11.12) can be represented graphically. The plot is shown in Fig. 11.10 for the particular case when the constant is equal to 0.8. Note that the experimental data tend to lie in a very broad band, and that V_1^* can easily exceed unity.

(b) *The Kutateladze number correlation*. Alternative dimensionless velocities can be defined using the surface tension, σ, and not the tube diameter. These are the Kutateladze numbers, for the gas K_g and for the liquid K_1.

$$K_g = V_g \rho_g^{\frac{1}{2}} / [g\sigma(\rho_1 - \rho_g)]^{\frac{1}{4}} \qquad (11.15)$$

$$K_1 = V_1 \rho_1^{\frac{1}{2}} / [g\sigma(\rho_1 - \rho_g)]^{\frac{1}{4}} \qquad (11.16)$$

The most commonly quoted correlation of this type is that of Pushkina and Sorokin (1969)

$$K_g = 3.2. \qquad (11.17)$$

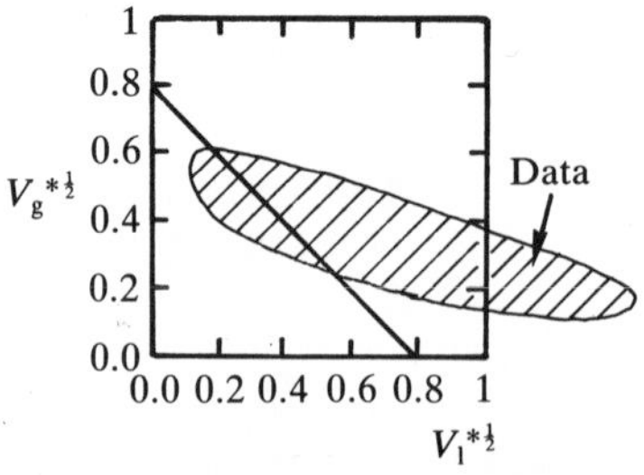

Fig. 11.10. Wallis-type flooding correlation.

It is sometimes said that to obtain good results use:

The Wallis-type correlation if the tube diameter is small (<50 mm), and

The Pushkina and Sorokin correlation if the tube diameter is large (>50 mm).

However, better overall results are obtained with a single, optimized correlation (see McQuillan and Whalley 1985)

$$K_g = 0.286 \, Bo^{0.26} Fr^{-0.22} \left[1 + \frac{\mu_1}{\mu_w}\right]^{-0.18}, \tag{11.18}$$

where:

Bo is the Bond number = $d^2 g(\rho_1 - \rho_g)/\sigma$; $\qquad(11.19)$

Fr is the Froude number = $V_1 \dfrac{\pi}{4} d \left[\dfrac{g(\rho_1 - \rho_g)^3}{\sigma^3}\right]^{\frac{1}{4}}$; $\qquad(11.20)$

μ_1 is the liquid viscosity (N s/m^2); and
μ_w is the water viscosity (at room temperature) = 10^{-3} N s/m^2.

Equation (11.18) was developed from a correlation originally devised by Alekseev *et al.* (1972) for flooding in packed beds. The use of the correlation of McQuillan and Whalley (1985) is given in Example 6 in Chapter 25.

11.8 A simple justification of K_g = constant

If we consider flooding to occur when a liquid drop can be supported by the gas flow then, writing a force balance on a drop of diameter D_d, the drag force F is given by

$$F = \frac{\pi D_d^3}{6}(\rho_1 - \rho_g)g. \tag{11.21}$$

The drag coefficient C_D is defined as

$$C_D = \frac{4F/\pi D_d^2}{\frac{1}{2}\rho_g u_g^2}, \tag{11.22}$$

and so

$$F = \frac{1}{8}\rho_g u_g^2 C_D \pi D_d^2 \tag{11.23}$$

or

$$u_g^2 = \frac{4D_d}{3} \frac{(\rho_1 - \rho_g)g}{\rho_g C_D}. \tag{11.24}$$

One way of correlating drop size D_d is to use a Weber number We

$$We = \rho_g u_g^2 D_d / \sigma \qquad (11.25)$$

or

$$D_d = \frac{We\,\sigma}{\rho_g u_g^2}. \qquad (11.26)$$

Then, eliminating D_d between eqn (11.24) and eqn (11.26) gives

$$u_g^4 = \frac{4\,We}{3C_D}\frac{\sigma(\rho_l - \rho_g)g}{\rho_g^2}, \qquad (11.27)$$

therefore

$$u_g = \left[\frac{4\,We}{3C_D}\right]^{\frac{1}{4}}\frac{[\sigma(\rho_l - \rho_g)g]^{\frac{1}{4}}}{\rho_g^{\frac{1}{2}}}, \qquad (11.28)$$

or

$$K_g = \left[\frac{4\,We}{3C_D}\right]^{\frac{1}{4}}. \qquad (11.29)$$

Moalem Maron and Dukler (1984) have suggested putting $We = 30$ and $C_D = 0.44$, giving $K_g = 3.1$, which is close to the previously quoted result of Pushkina and Sorokin. However, Hinze (1955) found that for a drop suddenly accelerated $We = 13$, giving $K_g = 2.5$. It is difficult to justify $We = 30$.

It is more difficult to justify the Wallis type of correlation (eqn (11.12)) on theoretical grounds, although Wallis (1969) gives a justification using mixing length theory. However another simple theory does give an interesting result.

11.9   A simple theory based on visualization results

Here, an attempt is made to calculate the gas velocity required to hold the liquid wave stationary. We are thus concerned with a force balance on the wave (see Fig. 11.11). The gas passes through a constriction formed by the wave. If there is no separation before the crest of the wave, then the pressure drop Δp across the wave is given by

$$\Delta p = \tfrac{1}{2}\rho_g u_g^2\left[1 - \frac{A}{A_1}\right]^2, \qquad (11.30)$$

where the areas A and A_1 (see Fig. 11.11), are given by

$$A = \frac{\pi d^2}{4} \qquad (11.31)$$

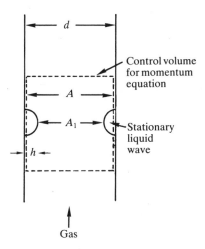

Fig. 11.11. Control volume for momentum equation calculation on a stationary wave.

and

$$A_1 = \frac{\pi}{4}(d - 2h)^2. \tag{11.32}$$

Then, if $h/d \ll 1$,

$$\Delta p = \tfrac{1}{2}\rho_g u_g^2 \frac{16h^2}{d^2}. \tag{11.33}$$

Now, using the momentum equation and the control volume shown in Fig. 11.11,

$$\Delta p A = (\rho_1 - \rho_g)V_{\text{wave}}g, \tag{11.34}$$

where V_{wave} is the volume of the wave which is given by

$$V_{\text{wave}} = \pi h^2 \left[\frac{\pi d}{2} - \frac{4h}{3}\right]. \tag{11.35}$$

For the case where $h/d \ll 1$,

$$V_{\text{wave}} = \frac{\pi^2 h^2 d}{2}. \tag{11.36}$$

Substituting eqns (11.33) and (11.36) into eqn (11.34) gives

$$\tfrac{1}{2}\rho_g u_g^2 \frac{16h^2}{d^2}\frac{\pi d^2}{4} = (\rho_1 - \rho_g)\frac{\pi^2 h^2}{2}dg \tag{11.37}$$

or

$$\frac{u_g^2 \rho_g}{d(\rho_1 - \rho_g)g} = \frac{\pi}{4}. \tag{11.38}$$

This is equivalent to

$$V_g^* = \left(\frac{\pi}{4}\right)^{\frac{1}{2}} = 0.89. \qquad (11.39)$$

The final result, eqn (11.39), lies near many experimental results and, in fact, describes the experimental data better than many more complicated theoretical models. It can be noted that the effects of the liquid film and the interfacial shear stress have been neglected.

References

ALEKSEEV, V. P., POBEREZKIN, A. E., and GERASIMOV, P. V. (1972). Determination of flooding rates in regular packings. *Heat Transfer Soviet Research* 4 (No. 6), 159–63.

HEWITT, G. F. (1982). Flow regimes. In *Handbook of multiphase systems* (ed. G. Hetsroni). McGraw-Hill, New York.

HEWITT, G. F., LACEY, P. M. C., and NICHOLLS, B. (1965). Transitions in film flow in a vertical tube. *Proc. Symp. on Two-Phase Flow, Exeter, UK.*

HINZE, J. O. (1955). Fundamentals of the hydrodynamic mechanism of slitting in dispersion process. *AIChE J.* 1, 289.

McQUILLAN, K. W. and WHALLEY, P. B. (1985). A comparison between flooding correlations and experimental flooding data, *Chem. Eng. Sci.* 40, 1425–40. (*AERE-R11267*, 1984.)

McQUILLAN, K. W., WHALLEY, P. B., and HEWITT, G. F. (1985). Flooding in vertical two-phase flow. *Int. J. Multiphase Flow* 11, 741–60.

MOALEM MARON, D. and DUKLER, A. E. (1984). Flooding and upward film flow in vertical tubes. II speculations on film flow mechanisms. *Int. J. Multiphase Flow* 10, 599–621.

PUSHKINA, O. L. and SOROKIN, Y. L. (1969), Breakdown of liquid film motion in vertical tubes. *Heat Transfer Soviet Research* 1 (no. 5), 56–64.

WALLIS, G. B. (1961). Flooding velocities for air and water in vertical tubes. *AEEW-R123.*

WALLIS, G. B. (1969). *One dimensional two-phase flow.* McGraw-Hill, New York.

WHALLEY, P. B. and McQUILLAN, K. W. (1985). Flooding in two-phase flow: the effect of tube length and artificial wave injection. *Physico-Chemical Hydrodynamics* 6, 3–21. (*AERE-R10883*, 1983.)

12

INSTABILITIES IN
TWO-PHASE FLOW

12.1 Introduction

Two-phase flows are prone to various types of instability (see Ishii 1982). These can be divided into two classes: static instability and dynamic instability.

A static instability is one where a small change in one of the independent variables leads to a large change in one of the dependent variables. Examples of static instability are listed below.

(*a*) *Flow regime transition.* A small change in the phase flow rates can make a large difference to the flow and hence, possibly, to the pressure drop or void fraction. Probably the clearest example of a flow pattern transition instability is flooding (see Chapter 11) where a very small change in, for example, gas velocity can radically change the pressure drop and the character of the flow.

(*b*) *The transition in the heat transfer characteristics of the flow at burnout or dryout.* A small change in flow rate can reduce the heat transfer coefficient by a factor of 100. This phenomenon is discussed in detail in Chapters 17 and 19.

(*c*) *The Ledinegg (1938) instability.* See § 12.2.

A dynamic instability is one which has an oscillatory nature. Examples of dynamic instability are as follows.

(*a*) *Chugging instability.* Chugging occurs particularly when liquid metals (for example, sodium) are boiled. The metal wets the surface very efficiently, and few nucleation sites are left. Consequently, the liquid can be superheated quite markedly before boiling starts. The liquid then vaporizes rapidly and, as the vapour occupies a much larger volume, the liquid tends to be thrown out of the channel. There follows a quiescent period while the channel refills with liquid.

(*b*) *Density wave instability.* This complex form of dynamic instability is discussed in § 12.3.

12.2 Ledinegg instability

This is the most interesting form of static instability. Consider the experimental situation illustrated in Fig. 12.1. Liquid is pumped up the

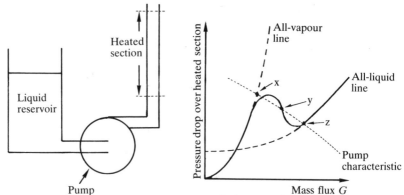

Fig. 12.1. Apparatus for investigating   Fig.   12.2. Ledinegg instability: possible
Ledinegg instability.   operating points.

heated tube from the reservoir. If the heat input rate to the tube is constant, then the pressure drop will vary with the mass flux in a complicated way. If there is single-phase liquid flow the pressure drop will be approximately proportional to G^2/ρ_l, and if there is single-phase vapour flow it will be approximately proportional to G^2/ρ_g. As the liquid vaporizes, the pressure drop can fall with increasing mass flux because the gravitational component of the total pressure drop is decreasing quickly, while the frictional pressure drop is increasing relatively slowly. The solid line in Fig. 12.2 represents, therefore, a typical variation of pressure drop with mass flux for an evaporating flow.

The other important pressure change in the system in Fig. 12.1 is that imposed by the pump. The pressure rise across the pump is not constant but varies with flow rate; this variation is known as the pump characteristic. To find the operating point for the system, the pump characteristic must also be plotted in Fig. 12.2. It can be seen that three intersections, and therefore three operating points, are possible: x, y and z. Here y is an unstable point and so the system will operate at x or at z. The unstable nature of point y can readily be seen if it is imagined that the system is operating at y and then the mass flux increases or decreases by a small amount. A random small decrease in the mass flux at y will cause an increase in the pressure drop but a smaller increase in the pressure provided by the pump. The flow will therefore decrease further until the point x is reached.

In a large boiler with many parallel channels between inlet and outlet headers, it would be possible to have some tubes operating at x and some at z. In an extreme case the flow at x could be so low that the quality is very high and the critical heat flux is thus exceeded. At z the flow might be so high that the flow remains as single-phase liquid flow and therefore

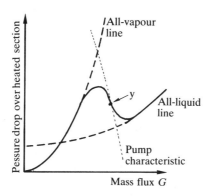

Fig. 12.3. Ledinegg instability: effect of making the pump characteristic line steeper.

the heat transfer is relatively poor. The situation can be improved by making the pump characteristic line steeper. There is now only one possible operating point, at y, and y is now a stable point (see Fig. 12.3).

12.3 Density wave instability

The most common, and most complicated, form of dynamic instability is a density wave instability. A typical boiling channel can be pictured as shown in Fig. 12.4.

If the pressures at A and B in the inlet and outlet headers are constant, then the flow through the boiling channel will be a solution of the relationship between the flow and the pressure drop. Static, that is steady-state, solutions have already been discussed as the Ledinegg instability problem (see § 12.2). However, there are further possible dynamic solutions. These arise because the various components of the total pressure drop are not necessarily in phase with each other. The

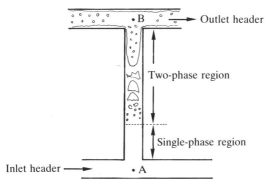

Fig. 12.4. Apparatus for invesigating density wave instability between constant pressure headers.

various components are:

(a) the pressure drop across an inlet restriction*;
(b) the frictional pressure drop in the single-phase region*;
(c) the gravitational pressure drop in the single-phase region*;
(d) the frictional pressure drop in the two-phase region;
(e) the gravitational pressure drop in the two-phase region;
(f) the accelerational pressure drop in the two-phase region; and
(g) the pressure drop across an outlet restriction.

Note that the components marked with an asterisk (*) are in phase with any inlet flow variation. If these components are large, then the flow will be stable. The other components are not in phase with any inlet flow variation. It is thus possible to have an oscillatory flow at the inlet superimposed on the steady flow so that the components of the pressure drop add up to the constant known pressure difference between A and B in Fig. 12.4. In order to stabilize a flow, it is necessary to make sure that:

(a) inlet pressure drop is large;
(b) single-phase length is large;
(c) exit quality is low; and
(d) exit pressure drop is low.

Also, an increase in the system pressure tends to make the system more stable. Many complicated computer programs and analyses have been performed for this problem of density wave instability (see, for example, Boure et al. 1973). If the system parameters, for example the void fraction variation with quality, pressure and flow rate, are known accurately, then the onset of the instability and its frequency can be predicted accurately. In a real boiling flow, if an instability is encountered, the usual way to eliminate it is to add an extra pressure drop in the single-phase liquid region. This is usually done by putting an orifice or valve into the channel.

References

Boure, J., Bergles, A. E., and Tong, L. S. (1973). Review of two-phase flow instability. *Nucl. Eng. Des.* **25**, 165–192.
Ishii, M. (1982). Two-phase flow instabilities. In *Handbook of multiphase systems* (ed. G. Hetsroni). McGraw-Hill, New York.
Ledinegg, M. (1983). Instabilität der Stömung bei Natürlichen und Zwangumlaut. *Warme* **61** (8), 891–8.

13

TWO-PHASE FLOW IN PIPE FITTINGS

13.1 Introduction

Two-phase flow in pipe fittings such as bends, enlargements, contractions, T junctions, and valves, is very complicated and has not been much studied. The overall subject can be divided into a number of smaller topics: flow patterns, pressure drop, and flow splitting. The information on each of these topics is, however, very incomplete and only a few general comments can be made.

13.2 Flow patterns

Flow patterns have been studied in only a few geometries, notably bends (usually in the form of coils) and changes of channel cross section.

It might be thought that, for flow in a bend, liquid would tend to be thrown to the outside of the bend, as shown in Fig. 13.1a. However, this is not necessarily the case. The force per unit volume on each phase is $\rho_i u_i^2 / R$, where ρ_i and u_i are the phase density and the actual phase velocity, and R is the radius of curvature. Thus, the phase with the larger value of $\rho_i u_i^2$ will tend to move to the outside of the bend. The following is normally found.

(a) At high pressures $\rho_g u_g^2 < \rho_l u_l^2$, and so the liquid moves to the outside of the bend. This is the normal behaviour illustrated in Fig. 13.1(a).

(b) At low pressures $\rho_g u_g^2 > \rho_l u_l^2$, and so the gas moves to the outside of the bend. This is the inverted type of behaviour found by Banerjee et $al.$ (1967) and by Whalley (1980), and illustrated in Fig. 13.1(b).

It is also known that bends are important in modifying the annular flow pattern in other ways.

(a) The entrained liquid drops are flung to the outside wall of the bend because they are travelling at approximately the same velocity as the gas. The entrained liquid fraction in a helically coiled tube, or just after a bend, therefore tends to be low. This fact implies that the critical heat flux (see Chapters 17 and 19) is higher than for straight tubes.

(b) In the sharp 180° bends of a serpentine boiler the disturbance

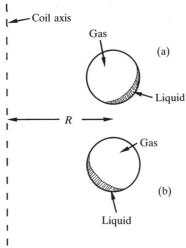

Fig. 13.1. Liquid film behaviour in a bend or coil: (a) normal behaviour, and (b) inverted behaviour.

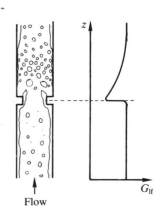

Fig. 13.2. Annular flow in a tube with an orifice plate.

waves on the liquid film are destroyed, but form again a short distance downstream.

Annular flow has also been studied during the flow across an orifice plate, see Fig. 13.2 (McQuillan and Whalley 1984). The effect of the orifice plate was to greatly increase the entrained liquid fraction just downstream of the orifice plate.

The extra liquid drops quickly deposit again and the liquid film flow regains its original value. Just as in the 180° bend, the orifice plate destroys the disturbance waves, but they form again downstream of the orifice plate.

13.3 Pressure drop

Various very specialized methods, often based on very little data, have been developed for particular geometrries. In many cases it is sufficient to use the homogeneous or the separated flow models. Therefore, if Δp_{lo} (N/m^2) is the pressure drop across the fitting when all the mass flow is assumed to be liquid, then the pressure drop Δp (N/m^2) in the two-phase flow is given by

$$\Delta p = \Delta p_{\text{lo}} \phi_{\text{lo}}^2; \tag{13.1}$$

and for homogeneous flow, if the friction factor can be assumed to be constant,

$$\phi_{lo}^2 = \frac{x\rho_1}{\rho_g} + (1-x). \tag{13.2}$$

For separated flow the methods for pipe flow in Chapter 6 can be used. One equation which is used particularly when accelerational effects are dominant is

$$\phi_{lo}^2 = \frac{(1-x)^2}{1-\alpha} + \frac{\rho_1 x^2}{\rho_g \alpha}. \tag{13.3}$$

A useful summary of two-phase pressure in pipe fittings is given by Collier (1981).

13.4 Flow split at T-junctions

This problem is illustrated in Fig. 13.3: if the flow at inlet 1 is known, then can the flow at outlets 2 and 3 be determined?

Denote the mass flux and quality at each point by G_1, x_1, etc. If G_1, x_1 and G_2 are known, then there are three unknowns: x_2, G_3, and x_3. There are two obvious equations available.

(a) Using a total mass balance G_3 can be found.

(b) Using a mass balance on one of the phases (the gas is the easiest), x_2 and x_3 can be related. However, x_2 and x_3 cannot be found independently.

Note that the momentum equation is no assistance because there is an unknown force exerted by the T-junction on the fluid. There is no other equation which is obviously applicable to the system, and therefore the problem cannot be solved theoretically. Experimental results are, however, available.

One method of expressing these results is by plotting the data in the form shown in Fig. 13.4. The 45° line represents a perfect flow split, that is where there are equal qualities in each outlet tube. Each outlet tube is then receiving a true sample of the inlet mixture. It is evident that the

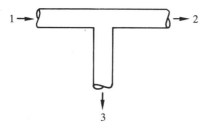

Fig. 13.3. T-junction: schematic diagram.

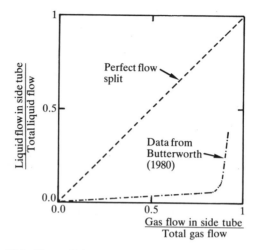

Fig. 13.4. Flow split in a T-junction (from Butterworth 1980).

flow split can, in the rather extreme case shown in Fig. 13.4, be very unequal. In this case the gas is tending to enter the side tube preferentially. This, however, is not always the case: sometimes the liquid preferentially enters the side tube. The behaviour of the T-junction has been found to be heavily influenced by the flow pattern in the main tube (Azzopardi and Whalley 1982).

In annular flow some progress has been made by picturing the split as shown in Fig. 13.5. The gas entering the side tube is imagined to come from the shaded area. The liquid entering the side tube is assumed to come from that part of the liquid film which is flowing in the shaded area. The entrained liquid drops are assumed not to enter the side tube. This is because their momentum is relatively large and, unlike the gas (which has a low density) or the liquid film (which has a low velocity), they are difficult to divert into the side tube.

It should be evident that splitting a two-phase flow is not good practice and should be avoided whenever possible. If it has to be done then the following rules should be followed.

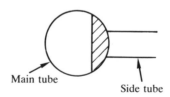

Fig. 13.5. Possible split in annular flow entering a T-junction.

(*a*) the best method is to separate the gas and liquid, for example in a cyclone, then to split the single-phase flows, and to recombine them subsequently to form the outlet two-phase flow. This, of course, is not always practicable due to the cost and size of the separation equipment required.

(*b*) If the two-phase flow is split without prior separation, then any design should take into account the fact that the split might be as bad as is possible. For example, in a two-pass heat exchanger then the heat exchanger should be designed so that it will work if all the liquid enters some tubes, and all the gas enters the remaining tubes.

References

AZZOPARDI, B. J. and WHALLEY, P. B. (1982). The effect of flow patterns on two-phase flow in a "T" junction. *Int. J. Multiphase Flow* **8,** 491–507.

BANERJEE, S., RHODES, E., and SCOTT, D. S. (1967). Film inversion of cocurrent two-phase flow in helical coils. *AIChE J* **13,** 189–91.

BUTTERWORTH, D. (1980). Unresolved problems in heat exchanger design. *Instn of Chem. Engnrs Symp. Ser.* **No. 60,** 231–48.

COLLIER, J. G. (1981). *Convective boiling and condensation.* McGraw-Hill, New York.

MCQUILLAN, K. W. and WHALLEY, P. B. (1984). The effect of orifices on the liquid distribution in annular two-phase flow. *Int. J. Multiphase Flow* **10,** 721–9.

WHALLEY, P. B. (1980). Air–water two-phase flow in a helically coiled tube. *Int. J. Multiphase Flow* **6,** 345–56.

14

BOILING: INTRODUCTION AND POOL BOILING

14.1 Types of boiling

Boiling can be divided into categories according to the mechanism occurring, and according to the geometric situation. The three mechanisms of boiling are:

(*a*) nucleate boiling, where vapour bubbles are formed (usually at a solid surface);

(*b*) convective boiling, where the heat is conducted through a thin film of liquid—the liquid then evaporates at the vapour–liquid interface with no bubble formation; and

(*c*) film boiling, where the heated surface is blanketed by a film of vapour—the heat is conducted through the vapour, and the liquid vaporizes at the vapour–liquid interface.

The two main geometric situations are:

(*a*) pool boiling, where the boiling occurs at a heated surface in a pool of liquid which, apart from any convection induced by the boiling, is stagnant; and

(*b*) flow boiling, where the liquid is pumped through a heated channel, typically a tube.

Nucleate boiling and film boiling occur in both pool boiling and flow boiling but convective boiling occurs only in flow boiling.

First, we look at a classical pool boiling experiment which illustrates nucleate boiling and film boiling.

14.2 Pool boiling

In 1934 Nukiyama performed the experiment illustrated in Fig. 14.1. A platinum wire immersed in water was heated electrically. The current in the wire and voltage across the ends of the wire enable the power, and therefore the heat flux, to be calculated. Also, from the resistance of the wire the temperature of the wire can be found. The results are illustrated in Fig. 14.2. The curve of the heat flux ϕ against the temperature difference ΔT_{sat} (which is the wall temperature minus the liquid satura-

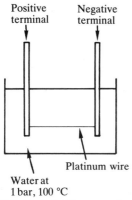

Fig. 14.1. Pool boiling experiment in which the heat flux is controlled.

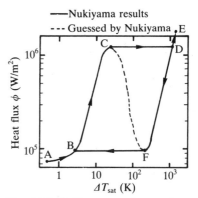

Fig. 14.2. Boiling curve from a heat-flux-controlled surface.

tion temperature) is the boiling curve. The regions of the curve are:

A to B   natural-convection single-phase liquid—there is no boiling in this region;
B to C   nucleate boiling; and
F to D to E   film boiling.

The important points on the curve are:

B   the onset of nucleate boiling;
C   The burnout point, where the heat flux is equal to the critical heat flux; and
F   The minimum film boiling point.

Note that when, as in the case of the apparatus shown in Fig. 14.1, the heat flux is controlled there is a hysteresis loop. The complete curve (shown by a dotted line in Fig. 14.2) was guessed by Nukiyama. The complete curve can be obtained by controlling the temperature rather than the heat flux. This can be done by, for example, heating the heat transfer surface by means of a hot liquid, as illustrated in Fig. 14.3 (see Bennett et al. 1967). The heat flux can be calculated by measuring the temperature drop in the hot liquid. The boiling curve obtained is shown in Fig. 14.4. It can be noted that in this case there is no hysteresis. The part of the boiling curve between C and F is known as the transition boiling region.

The complete boiling curve can be calculated using equations in subsequent chapters (see Example 7 in Chapter 25).

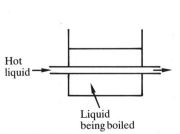

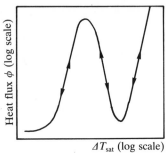

Fig. 14.3. Pool boiling experiment in which the wall temperature is controlled.

Fig. 14.4. Boiling curve from a temperature-controlled surface.

14.3 Visualization of events along the boiling curve

Direct visual and photographic evidence shows the following.

(a) The nucleate boiling region (BC in Fig. 14.2) consists of two parts:

   (i) the isolated bubble region, where bubbles behave independently, (as illustrated in Fig. 14.5(a)); and
   (ii) the slugs and columns region, where the bubbles start to merge and to depart from the heated surface by means of jets which then form large bubbles, or slugs, above the surface (see Fig. 14.5(b)).

(b) The film boiling region (FDE in Fig. 14.2) is where the heated surface is covered with a layer of vapour (see Fig. 14.5(c)). The liquid is not in contact with the heated surface. The vapour surface is unstable, and bubbles are released from it into the liquid.

(c) The transition boiling region (FC in Fig. 14.2) is a complex region where parts of the surface are in film boiling regime and parts in the nucleate boiling regime of the slugs and columns type.

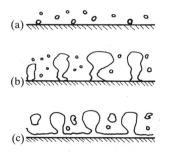

Fig. 14.5. Visualization results in nucleate and film boiling.

14.4 Bubble growth in nucleate boiling

Nucleate boiling can occur when the bulk liquid is saturated (when it is at its boiling temperature) or when it is subcooled (when it is below its boiling temperature). One difference is in what happens to the growing bubbles at the heating surface. In saturated liquid a bubble grows and then, aided by buoyancy, leaves the surface. As it leaves, fresh liquid flows towards the surface. Another bubble then begins to grow at the same point. In subcooled liquid the bubble grows and reaches out into the relatively cool liquid: thus the vapour begins to condense, and in doing so causes the liquid temperature to rise slightly. The bubble can collapse completely and, once again, new, cool liquid flows into the area near the wall; the process of bubble growth can then start again.

The collapsing of bubbles in subcooled liquid is responsible for the characteristic 'singing' sound of a kettle when relatively cool water is being heated up to its boiling point. It can also be noted that the bubble is only in contact with a very small area of the heated wall. there is normally no large, dry patch underneath the bubble.

Bubble growth is controlled by two factors.

(*a*) *The inertia of the liquid.* This is relevant at short times after the bubble is formed, and while the inertia is the controlling effect the bubble radius is proportional to time elapsed since the bubble was formed.

(*b*) *Thermal diffusion through a boundary layer around the bubble.* The latent heat for the evaporation must be supplied from the hot liquid layer around the growing bubble. While the thermal diffusion effect is controlling the radius it is proportional to the square root of time elapsed since bubble formation. Thermal diffusion becomes the controlling effect at relatively long times after the bubble starts its growth.

Thus the bubble grows in the manner illustrated in Fig. 14.6. At short times the bubble radius is proportional to t (inertia-controlled). At long

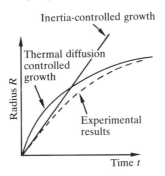

Fig. 14.6. Bubble growth as a function of time.

times the bubble radius is proportional to $t^{\frac{1}{2}}$ (thermal-diffusion-controlled).

14.5 Bubble nucleation

The process of bubble formation is known as nucleation. In many ways it is analogous to crystal nucleation. A salt solution can be cooled until it becomes supersaturated, but before a crystal can start growing it must have something to start growing upon: a rough surface, dirt particles, or a small crystal. Bubble nucleation of dissolved gas to form gas bubbles in a liquid can be seen in a carbonated drink in a glass. Careful observation will show that the gas bubbles:

(a) are formed at a surface on the glass container;

(b) rise in a chain of bubbles originating from the same spot on the surface; and

(c) often originate from the same spot if the glass is emptied and refilled.

If the liquid is clean (that is it does not contain dirt particles) the bubbles are not observed to arise from a spot inside the bulk of the liquid, away from the surface.

The surface properties are obviously important in all these nucleation phenomena: crystallization, gas bubble formation and nucleate boiling. The surface, when viewed under a high-power microscope, is not smooth: it contains pits and cracks in a complex pattern. Smoother and smoother surface finishes only make the size scale of the pits and cracks smaller and smaller. A very enlarged cross section through the surface might be as shown in Fig. 14.7. The cracks and crevices do not, of themselves, constitute nucleation sites for the bubbles: they must also contain pockets of gas, probably air trapped when the vessel was filled with the liquid. It is from these pockets of trapped air that the vapour bubbles begin to grow during nucleate boiling.

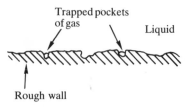

Fig. 14.7. Enlarged view of a boiling surface.

14.6 Types of nucleation

The usual type of nucleation is that described in the previous section: nucleation at a solid surface. This is called heterogeneous nucleation because a solid and a liquid are involved. In extreme cases, where there are no nucleation sites on the vessel walls, homogeneous nucleation may occur. This is nucleation which occurs in the bulk liquid away from the walls of the vessel. Such an extreme situation could be created by the apparatus shown in Fig. 14.8. Here, water is heated while contained in a rotating 'dish' of mercury. If the water and the mercury are clean, there will be no nucleation sites available for bubble growth. The temperature of the water, which is at a pressure of 1 bar, can be raised far above 100 °C before boiling starts.

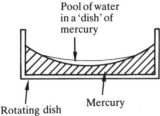

Fig. 14.8. Apparatus for a homogeneous nucleation experiment.

Homogeneous nucleation occurs when, by randomly occurring processes, a number of energetic molecules come together to form a small, very localized, 'vapour' region. This vapour region can then grow rapidly. Once homogeneous nucleation begins the vapour bubbles grow at an explosive rate.

For many organic substances, at 1 bar, homogeneous nucleation occurs (see Blander and Katz 1975) at about $0.89\ T_c$, where T_c is the critical temperature (K). The homogeneous nucleation temperature increases as the pressure increases, and becomes equal to the critical temperature when the pressure reaches the critical pressure. For water at 1 bar this rule does not work too well, and the homogeneous nucleation temperature is about 321 °C ($= 0.92\ T_c$). The latent heat necessary to fuel the large bubble growth rate is supplied from the very hot liquid.

Homogeneous nucleation can be important in some clean systems, using organic liquids, which are operated close to the critical pressure. However, under these conditions the vaporization is much more gentle and less explosive.

14.7 Heterogeneous nucleation in pool boiling

The ideal cavity is pictured (see Fig. 14.9) as being conical and having a circular opening. As the bubble grows its radius of curvature changes.

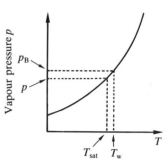

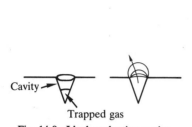

Fig. 14.9. Ideal nucleation cavity.   Fig. 14.10. Vapour pressure curve: superheat required for nucleation.

The minimum radius of curvature of the bubble occurs when the bubble forms a hemisphere at the cavity mouth. The radius of curvature is then, of course, equal to the radius R of the cavity opening.

If the pressure inside the bubble is p_B (N/m^2), then

$$p_B = p + \frac{2\sigma}{r}, \tag{14.1}$$

where p is the pressure in the liquid (N/m^2), and r is the bubble radius (m). Now, p_B is a maximum when $r = R$ (the cavity radius). The wall temperature T_w must be high enough to vaporize the liquid at a pressure of p_B (see Fig. 14.10). So, for the bubble to grow,

$$T_w > T_{sat} + \frac{dT}{dp}(p_B - p) \tag{14.2}$$

The slope of the vapour pressure curve can be found from the Clausius–Clapeyron equation

$$\frac{dp}{dT} = \frac{\lambda}{(v_g - v_l)T_{sat}}, \tag{14.3}$$

Where λ is the latent heat of vaporization (J/kg), T_{sat} is the saturation temperature (K), and v_g and v_l are the specific volumes of the gas and the liquid (m^3/kg). Then if $v_g \gg v_l$ and, since $v_g = 1/\rho_g$, then

$$\frac{dT}{dp} = \frac{T_{sat}}{\lambda \rho_g}, \tag{14.4}$$

and the inequality (14.2) becomes

$$T_w > T_{sat} + \frac{2\sigma}{R}\frac{T_{sat}}{\lambda \rho_g}. \tag{14.5}$$

If ΔT_{sat} is the value of $(T_w - T_{\text{sat}})$ at which nucleation starts, then the cavity radius is given by

$$R = \frac{2\sigma T_{\text{sat}}}{\rho_g \lambda \Delta T_{\text{sat}}}. \tag{14.6}$$

For water at 1 bar, ΔT_{sat} is commonly about 5 K so, putting

$T_{\text{sat}} = 373$ K,
$\sigma = 0.059$ N/m,
$\lambda = 2.256 \times 10^6$ J/kg, and
$\rho_g = 0.598$ kg/m³,

then $R = 6.5 \times 10^{-6}$ m $= 6.5$ μm.

Thus typical cavity sizes are in the micron range. If the cavity size is known, then clearly the wall superheat ΔT_{sat} required to start nucleate boiling can be calculated. Real surfaces, of course, can contain a range of cavity sizes. As the wall superheat ΔT_{sat} is increased, cavities of a smaller and smaller radius are able to become active and initiate nucleation.

14.8 Heat transfer in nucleate boiling

It is evident that it will be difficult to obtain any general theoretical method of calculating heat transfer coefficients in nucleate boiling. This is because the boiling occurs at nucleate sites, and the number of sites is very dependent upon:

(a) the physical condition and preparation of the surface; and
(b) how well the liquid wets the surface and how efficiently the liquid displaces air from the cavities.

Certainly, equations of the type

$$\phi \propto \Delta T_{\text{sat}}^{1.2} n^{0.33}, \tag{14.7}$$

where n is the nucleation site density (number per unit area), have been produced. However, this is not of much practical use if the variation of n with ΔT_{sat} is not known. Disregarding n, and simply looking at the variation of heat flux with temperature difference, it has been found many times that

$$\phi \propto \Delta T_{\text{sat}}^a, \tag{14.8}$$

where $a = 3$ to 3.33.

Since we can also write

$$\phi = h\,\Delta T_{\text{sat}}, \tag{14.9}$$

then

$$h \propto \Delta T_{\text{sat}}^{a-1} \qquad (14.10)$$

and

$$h \propto \phi^{a-1/a}, \qquad (14.11)$$

where

$$\frac{a-1}{a} = \frac{2}{3} \quad \text{when} \quad a = 3$$

and

$$\frac{a-1}{a} = 0.7 \quad \text{when} \quad a = 3.33.$$

The heat transfer coefficient $h(\text{W/m}^2\,\text{K})$ is high in nucleate boiling (typically $10\,\text{kW/m}^2\,\text{K}$). It is slightly higher for water, and lower for organic liquids. Nucleate boiling is thus a very efficient heat-transfer system.

Two practical approaches to the calculation of nucleate-boiling heat transfer are therefore possible.

(a) Attempt to take some account of the surface effects.

(b) Ignore all surface effects and produce a method which gives a typical heat-transfer coefficient at the particular heat flux for the fluid.

Rohsenow (1952) included some effects of the surface. He argued that in single-phase convective heat transfer

$$Nu = f(Re, Pr), \qquad (14.12)$$

where:

$$Nu = \text{Nusselt number} \quad = hL/k; \qquad (14.13)$$

$$Re = \text{Reynolds number} = uL\rho/\mu; \text{ and} \qquad (14.14)$$

$$Pr = \text{Prandtl number} \quad = \mu C_{\text{p}}/k. \qquad (14.15)$$

Taking the physical properties as those of the liquid, there are then two problems in trying to use this type of relation for boiling: what is the velocity u and what is the length scale L?

(a) The velocity is taken as the liquid velocity in towards the surface which is to supply the vapour which is being produced, so

$$u = \phi/\lambda\rho_1. \qquad (14.16)$$

(b) The length scale is taken to be

$$L = [\sigma/g(\rho_1 - \rho_g)]^{\frac{1}{2}}. \qquad (14.17)$$

This is related to the most unstable wave on a liquid–vapour interface (see Chapter 15 on critical heat flux in pool boiling).

Hence

$$Nu = \frac{h}{k_1}\left[\frac{\sigma}{g(\rho_1 - \rho_g)}\right]^{\frac{1}{2}}, \tag{14.18}$$

$$Re = \frac{\phi}{\lambda\rho_1}\left[\frac{\sigma}{g(\rho_1 - \rho_g)}\right]^{\frac{1}{2}}\frac{\rho_1}{\mu_1}, \tag{14.19}$$

and

$$Pr = \frac{\mu_1 C_{pl}}{k_1}. \tag{14.20}$$

Rohsenow then correlated Nu with Re and Pr, so that

$$Nu = \frac{1}{C_{sf}} Re^{1-n}Pr^{-m} \tag{14.21}$$

or, rearranging,

$$\frac{C_{pl}\Delta T_{sat}}{\lambda} = C_{sf}\left[\frac{\phi}{\mu_1\lambda}\left(\frac{\sigma}{g(\rho_1 - \rho_g)}\right)^{\frac{1}{2}}\right]^{n}\left(\frac{\mu_1 C_{pl}}{k_1}\right)^{1+m}, \tag{14.22}$$

where, commonly, $n = 0.33$ and $m = 0.7$. Note that $n = 0.33$ is equivalent to $a = 3$ in eqn (14.8). C_{sf} is the surface–fluid constant. It depends on both the surface and the fluid. Typical values are between 0.0025 and 0.015. Note that, for a given ΔT_{sat}, the heat flux is proportional to C_{sf}^{-3}. Since C_{sf} can vary by a factor of 10, the heat flux can vary by a factor of 1000.

The heat flux at the onset of nucleate boiling can be found by using eqn (14.5) to find ΔT_{sat} when nucleate boiling starts, and then by using eqn (14.22) to find the corresponding heat flux.

Often the surface effects are unknown and C_{sf} is not known, so Rohsenow's correlation is no help. This has led to the development of methods which ignore all surface effects, and are thus relatively inaccurate.

A typical overall blanket correlation is that of Mostinski (see Mostinski 1963; and Starczewski 1965).

$$h = 0.106p_c^{0.69}\phi^{0.7}f(p_R), \tag{14.23}$$

where h is the pool-boiling heat-transfer coefficient (W/m² k), ϕ is the heat flux (W/m²), p_c is the critical pressure (bar), $f(p_R)$ is a function of the reduced pressure p/p_c, and

$$f(p_R) = (1.8p_R^{0.17} + 4p_R^{1.2} + 10p_R^{10}). \tag{14.24}$$

Equation (14.23) is a dimensional equation and so the units given above must be used. Note that $h \propto \phi^{0.7}$, and that all the other physical properties and their variations with pressure are 'buried' in p_c and p_R. h increases monotonically with p_R, and rises rapidly as the critical pressure $(p_R = 1)$ is reached. This type of equation is known to work reasonably well, but why? Cooper (1984) has recently explained that properties like density, surface tension, and latent heat can be expressed as functions of the reduced variables such as p_R. Moreover, Cooper has derived a new correlation

$$h = A p_R^{(0.12 - \log_{10}\varepsilon)} (-\log_{10} p_R)^{-0.55} M^{-0.5} \phi^{\frac{2}{3}}, \qquad (14.25)$$

where M is the molecular weight (for example for water $M = 18$) and A is a constant. A best-fit line through the data gave $A = 55$, and a conservative value of A is ~ 30–40. ε is the surface roughness in microns. The units of h are $(\mathrm{W/m^2\,K})$ and the units of ϕ are $(\mathrm{W/m^2})$. Equation (14.25) is also a dimensional equation, and so the units given above must be used. Like the Mostinski equation this equation gives values of the heat transfer coefficient which increase with pressure, but some attempt to include surface properties has been made by including the roughness ε. As the roughness increases, the heat transfer coefficient also increases.

An example of the use of eqn (14.25) is given in Example 7 in Chapter 25.

14.9 Effect of liquid velocity on nucleate boiling

So far only pure nucleate boiling has been discussed. What, however, is the effect of superimposing a liquid flow, as in Fig. 14.11(a)? Single-phase forced convection would give a constant heat-transfer coefficient for a given liquid velocity (line A) in Fig. 14.11(b), whereas nucleate boiling gives a much steeper line (line B). It seems (see Bergles and

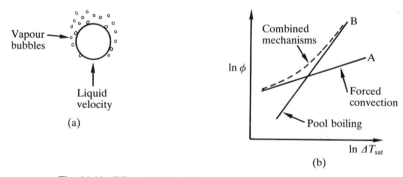

Fig. 14.11. Effect of a fixed liquid velocity on nucleate boiling.

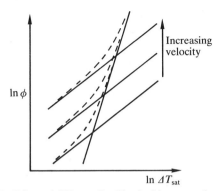

Fig. 14.12. Effect of different liquid velocities on nucleate boiling.

Rohsenow 1964; Kutateladze 1969) that the mechanism gradually changes from single-phase forced convection to nucleate boiling as the heater temperature is raised (see the dotted line in Fig. 14.11(b)).

At different liquid velocities, the single-phase forced convection heat-transfer coefficient increases with the velocity (see Fig. 14.12), but the nucleate boiling line is unaltered. So at high liquid velocities nucleate boiling only begins to dominate at a comparatively large value of ΔT_{sat}.

References

BENNETT, A. W., HEWITT, G. F., KEARSEY, H. A., and KEEYS, R. K. F. (1967). Heat transfer to steam–water mixtures flowing in uniformly heated tubes in which the critical heat flux has been exceeded. *AERE-R5373*.

BERGLES, A. E. and ROHSENOW, W. M. (1964). The determination of forced-convection surface boiling heat transfer. *J. Heat Transfer* **86**, 365–82.

BLANDER, R. M. and KATZ, J. L. (1975). Bubble nucleation in liquids. *AIChE J.* **21**, 833–43.

COOPER, M. G. (1984). Saturation nucleate pool boiling—a simple correlation. *1st UK National Heat Transfer Conference* (I. Chem. E. Symp. series No. 86) **2**, 785–93.

KUTATELADZE, S. S. (1969). Boiling heat transfer. *Int. J. Heat Mass Transfer* **4**, 31–45.

MOSTINSKI, I. L. (1963). Calculation of heat transfer and critical heat flux in boiling liquids based on the law of corresponding states. *Teploenergetika* **10**, (No. 4), 66–71.

NUKIYAMA, S. (1934). The maximum and minimum values of the heat transmitted from metal to boiling water at atmospheric pressure. Available as *Int. J. Heat Mass Transfer* (1966) **9**, 1419–33.

ROHSENOW, W. M. (1952). A method of correlating heat transfer data for surface boiling of liquids. *Trans. ASME*, **74**, 969.

STARCZEWSKI, J. (1965). Generalised design of evaporators—heat transfer to nucleate boiling liquids. *Brit. Chem. Eng.* **10**, 523–31.

15

CRITICAL HEAT FLUX IN POOL BOILING

15.1 Introduction

In pool boiling the critical heat flux occurs when the heated surface is covered with vapour bubbles, and these bubbles form a barrier to incoming liquid. For a heat-flux controlled surface the temperature rise when the critical heat flux is exceeded can be very large (sometimes more than 1000 K).

The steps in the classical theory of pool-boiling critical heat flux are (from Lienhard 1981):

(a) the instability of the vapour layer on the heated surfaces, and
(b) the instability of the vapour rising from the vapour layer through the liquid.

This is particularly interesting because it is an example of a comparatively simple theory which gives excellent results.

15.2 Instability of the vapour layer

A light fluid in a layer which has a heavy fluid on top of it is unstable. The layer breaks down by the formation of waves on its surface (Fig. 15.1). The wavelength of the most unstable wave (the one with the fastest growth rate) is λ_T. These waves are known as Taylor waves. For a horizontal layer as shown in Fig. 15.1 (see Bellman and Pennington 1954)

$$\lambda_T = C\left[\frac{\sigma}{(\rho_l - \rho_g)g}\right]^{\frac{1}{2}}. \tag{15.1}$$

For one-dimensional waves, as on a horizontal rod, $C = 2\pi\sqrt{3}$, and for two-dimensional waves, as on a horizontal flat plate, $C = 2\pi\sqrt{6}$. It can be noted that the length scale used in the development of Rohsenow's correlation for nucleate boiling (see § 14.8) was $[\sigma/(\rho_l - \rho_g)g]^{\frac{1}{2}}$.

From a horizontal plane surface, the vapour tends to arise as jets or columns arranged on a staggered square lattice, as shown in Fig. 15.2. The spacing of the jets is determined by the peaks in the unstable wave, so the separation of the peaks is λ_T. The two-dimensional result, as

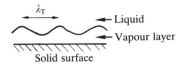

Fig. 15.1. Instability in the vapour layer on a flat plate.

shown in Fig. 15.2, is

$$\lambda_T = 2\pi\sqrt{6}\left[\frac{\sigma}{(\rho_1 - \rho_g)g}\right]^{\frac{1}{2}}. \tag{15.2}$$

The dimension x then in Fig. 15.2 is given by

$$x = \frac{\lambda_T}{\sqrt{2}} = 2\pi\sqrt{3}\left[\frac{\sigma}{(\rho_1 - \rho_g)g}\right]^{\frac{1}{2}}, \tag{15.3}$$

which is also the one-dimensional result. These results, then, describe the formation of the vapour jets at the peak amplitude position on the Taylor waves.

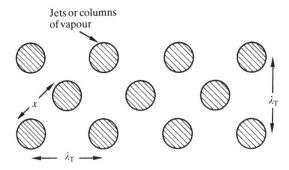

Fig. 15.2. Vapour jets arising from a flat plate.

15.3 Instability of the vapour jets

A parallel-sided jet is not stable in the configuration shown in Fig. 15.3(a). Consider the random thinning of the jet which is illustrated in Fig. 15.3(b).

By continuity, $u_2 > u_1$ and therefore, from Bernoulli's equation $p_2 < p_1$. If the jet is in equilibrium at 1, then the liquid pressure at 2 will push the 'neck' further in and disrupt the jet completely, thus breaking it up. This is a Kelvin–Helmholtz instability. This argument would imply that the vapour jet is always unstable, but the effects of surface tension, which has a stabilizing effect, have been neglected.

If the velocity of the vapour in the jet is u_g, then the dynamic pressure available is $\frac{1}{2}\rho_g u_g^2$. If the wavelength of the disturbance is λ_{KH} (as shown

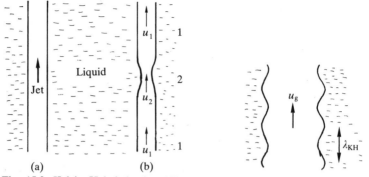

Fig. 15.3. Kelvin–Helmholtz instability   Fig. 15.4. Kelvin–Helmholtz wavelength.
in the vapour jet.

in Fig. 15.4), then the curvature of the surface (in the vertical direction) is proportional to λ_{KH}, and the pressure difference across this curved surface is proportional to σ/λ_{KH}. So when

$$\frac{\sigma}{\lambda_{KH}} \le A\tfrac{1}{2}\rho_g u_g^2, \tag{15.4}$$

where A is an unknown constant, the surface tension will no longer be able to stabilize the jet. Rearranging, the condition becomes

$$u_g \ge \left(\frac{2\sigma}{A\rho_g\lambda_{KH}}\right)^{\frac{1}{2}}. \tag{15.5}$$

In fact, a full analysis shows (see Turner 1973) that

$$u_g \ge \left(\frac{2\pi\sigma}{\rho_g\lambda_{KH}}\right)^{\frac{1}{2}}. \tag{15.6}$$

15.4   Calculation of the critical heat flux

Consider an area of the horizontal plate which supplies one jet (Fig. 15.5). This area is A_h and the jet cross-sectional area is A_j. Then, at the

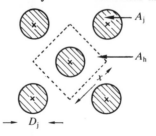

Fig. 15.5. Vapour jet diameter.

critical heat flux ϕ_c, the rate of heat supply to the area A_h is

$$\phi_c A_h = \lambda \rho_g u_g A_j, \tag{15.7}$$

and so

$$\phi_c = \lambda \rho_g u_g \frac{A_j}{A_h}. \tag{15.8}$$

Now, see Fig. 15.5,

$$A_h = x^2, \tag{15.9}$$

and D_j was first guessed, and then confirmed to be $x/2$, therefore $A_j = \pi(x/4)^2$ and

$$\frac{A_j}{A_h} = \frac{\pi \left(\frac{x}{4}\right)^2}{x^2} = \frac{\pi}{16}. \tag{15.10}$$

Substituting eqn (5.10) into eqn (15.8),

$$\phi_c = \lambda \rho_g u_g \frac{\pi}{16}, \tag{15.11}$$

where, at the critical condition (from eqn (15.6))

$$u_g = \left(\frac{2\pi\sigma}{\rho_g \lambda_{KH}}\right)^{\frac{1}{2}}. \tag{15.12}$$

Then, making the assumption that

$$\lambda_{KH} = x, \tag{15.13}$$

and that

$$x = 2\pi\sqrt{3}\left[\frac{\sigma}{(\rho_1 - \rho_g)g}\right]^{\frac{1}{2}}, \tag{15.14}$$

and then, substituting eqns (15.12–14) into eqn (15.11), we get

$$\phi_c = \lambda \rho_g \frac{\pi}{16} \left(\frac{2\pi\sigma}{\rho_g}\right)^{\frac{1}{2}} \left[\frac{(\rho_1 - \rho_g)g}{\sigma}\right]^{\frac{1}{4}} \left(\frac{1}{2\pi\sqrt{3}}\right)^{\frac{1}{2}} \tag{15.15}$$

or, simplifying,

$$\phi_c = \frac{\pi}{16 \times 3^{\frac{1}{4}}} \lambda \rho_g^{\frac{1}{2}} [\sigma(\rho_1 - \rho_g)g]^{\frac{1}{4}}. \tag{15.16}$$

The numerical constant $\pi/(16 \times 3^{\frac{1}{4}})$ has the value of 0.149.

This formula for critical heat flux has been derived in a number of different ways; each time the same formula has been obtained, although with a different constant. Zuber's (1959) original analysis gave a value for the constant of $\pi/24 = 0.131$. The assumption inherent in eqn (15.13), that the Kelvin–Helmholtz wavelength is the same as the Taylor one-dimensional wavelength can, in fact, be proved to be true.

It will be evident that the equation for critical heat flux is reminiscent of the Kutateladze number criterion that flooding occurs when

$$K_g = V_g \rho_g^{\frac{1}{2}} [g\sigma(\rho_1 - \rho_g)]^{-\frac{1}{4}} = \text{constant}, \qquad (15.17)$$

where V_g is the superficial vapour velocity above the plate (m/s). If now we put

$$\phi_c = \lambda \rho_g V_g, \qquad (15.18)$$

then

$$\phi_c = K_g \lambda \rho_g^{\frac{1}{2}} [\sigma(\rho_1 - \rho_g)g]^{\frac{1}{4}}. \qquad (15.19)$$

Kutateladze (1948) performed experiments to determine K_g.

15.5 Experimental results

For flat, horizontal plates the equation

$$\phi_c = 0.149 \lambda \rho_g^{\frac{1}{2}} [\sigma(\rho_1 - \rho_g)g]^{\frac{1}{4}} \qquad (15.20)$$

works very well as long as two conditions are satisfied (see Lienhard and Dhir 1973).

(a) Liquid is prevented from entering around the sides of the plate, as it does in Fig. 15.6(a). This can be ensured by adding sides to the flat plate (see Fig. 15.6(b)).

(b) The test section should be reasonably large. If the test section dimensions become small then the number of jets to be fitted in becomes important. This makes ϕ_c vary substantially from the predictions of eqn (15.20) when the length is less than $3x$. This is illustrated in Fig. 15.7.

Eqn (15.20) is used in Example 7 in Chapter 25 as part of the calculation of the complete pool-boiling curve.

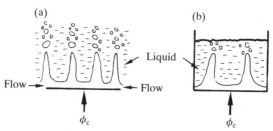

Fig. 15.6. Critical heat flux on a flat plate: conditions for theory to work well.

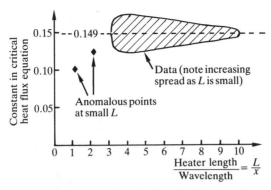

Fig. 15.7. Critical heat flux on a flat, horizontal plate: experiment and theory.

15.6 Critical heat flux for boiling outside horizontal cylinders

Horizontal cylinders are an important geometry. The basic theory works well, although the numerical constant in eqn (15.20) needs some adjustment because of this geometry. It is found that for large cylinder radii

$$\phi_c = 0.116\lambda\rho_g^{\frac{1}{2}}[\sigma(\rho_l - \rho_g)g]^{\frac{1}{4}}. \tag{15.21}$$

For small radii the constant is larger, and is illustrated in Fig. 15.8. The constant can be expressed as (see Sun and Lienhard 1970)

$$\text{numerical constant} = 0.116 + 0.3e^{-3.44R'^{\frac{1}{2}}}, \tag{15.22}$$

where the dimensionless radius R' is defined as

$$R' = R\left[\frac{\sigma}{g(\rho_l - \rho_g)}\right]^{-\frac{1}{2}} \tag{15.23}$$

and R is the actual radius of the cylinder (m). If $R' > 1$, then the cylinder

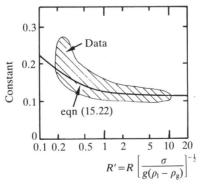

Fig. 15.8. Critical heat flux for boiling on a cylinder: variation with cylinder radius.

is effectively large. For water at 1 bar, a dimensionless radius of unity corresponds to an actual radius of 2.5 mm or a diameter of 5 mm. Thus, for tubes of industrial size, the numerical constant in eqn (15.21) can be taken to be 0.116.

15.7 Variation of critical heat flux with pressure

As the system pressure rises:

λ falls slowly at first and then falls steeply as the critical point is approached,
ρ_g increases monotonically,
σ falls monotonically, and
$\rho_g - \rho_g$ falls monotonically.

For steam–water, values of the critical heat flux for a flat, horizontal plate calculated from eqn (15.20) are given in Table 15.1.

TABLE 15.1 *Values of critical heat for a flat, horizontal plate using steam–water*

p (bar)	ϕ_c (MW/m^2)
0.01	0.168
0.1	0.471
1	1.25
10	2.97
30	4.03
50	4.38
70	4.45
90	4.34
100	4.10
150	3.27
221	0
(p_c)	

Note that the maximum critical heat flux occurs at about 70 bar. The variation with pressure is illustrated in Fig. 15.9. This can be compared with the flow-boiling critical heat flux variation with pressure for water

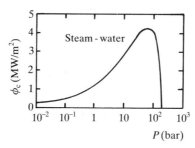

Fig. 15.9. Variation of critical heat flux on a flat, horizontal plate with pressure for steam–water.

given in Chapter 17. For both pool boiling and flow boiling the maximum critical heat flux occurs at about 70 bar.

15.8 Effect of liquid subcooling

All the previous results have been for a saturated liquid. If the liquid is cooled an amount ΔT_{sub}(K) below its saturation temperature, then the critical heat flux increases, and the increase is roughly proportional to ΔT_{sub}. One suggestion concerning this has been that made by Ivey and Morris (1962)

$$\phi_{c,sub} = \phi_{c,sat}\left[1 + 0.1\left(\frac{\rho_l}{\rho_g}\right)^{\frac{3}{4}}\frac{C_{pl}\Delta T_{sub}}{\lambda}\right], \tag{15.24}$$

$\phi_{c,sub}$ is the critical heat flux (W/m^2) when the liquid is cooled ΔT_{sub} below its saturation temperature and $\phi_{c,sat}$ is the critical heat flux (W/m^2) in saturated liquid. Substitution of typical values into eqn (15.24) show that even a moderate degree of subcooling increases the critical heat flux substantially.

15.9 Effect of liquid velocity

For small cylinders, of diameter less than 1 mm (effectively thin wires), the presence of an upward liquid velocity replaces the jet structure by a two-dimensional sheet, as shown in Fig. 15.10 (see Lienhard and Eichhorn 1976).

When the liquid velocity is sufficient to form the two-dimensional wake structure, then the critical heat flux is increased by the presence of the liquid velocity. At high liquid velocities the critical heat flux is proportional to the liquid velocity.

For larger cylinders, of diameter greater than about 15 mm (that is tubes, see McKee and Bell 1969), there is less evidence, but the critical heat flux does not seem to be greatly affected by the liquid velocity.

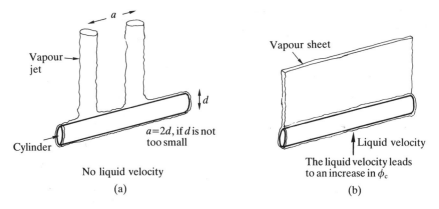

Fig. 15.10. Change in flow structure as a result of an imposed upward liquid velocity. (a) Flow structure with no imposed upward liquid velocity. (b) Flow structure with imposed upward liquid velocity.

References

BELLMAN, R. and PENNINGTON, R. H. (1954). Effects of surface tension and viscosity on Taylor instability. *Quart. J. Appl. Math.* **12**, 151.

IVEY, H. J. and MORRIS, D. J. (1962). On the relevance of the vapour–liquid exchange mechanism for subcooled boiling heat transfer at high pressure. *AEEW-R137.*

KUTATELADZE, S. S. (1948). On the transition to film boiling under natural convection, *Kotloturbostroenie* **3**, 10.

LIENHARD, J. H. (1981). *A heat transfer text book.* Prentice Hall, Englewood Cliffs, New Jersey.

LIENHARD, J. H. and DHIR, V. K. (1973). Extended hydrodynamic theory of the peak and minimum pool boiling heat fluxes. *NASA CR-2270.*

LIENHARD, J. H. and EICHHORN, R. (1976). Peak boiling heat flux of cylinder in a cross flow. *Int. J. Heat Mass Transfer* **19**, 1135–41.

McKEE, H. R. and BELL, K. J. (1969). Forced convection boiling from a cylinder normal to the flow. *Chem. Eng. Prog. Symp. Ser.* **65**, (No. 92) 222–30.

SUN, K. H. and LIENHARD, J. H. (1970). The peak pool boiling heat flux on horizontal cylinders. *Int. J. Heat Mass Transfer* **13**, 1425–39.

TURNER, J. S. (1973). Buoyancy effects in fluids. Cambridge University Press.

ZUBER, N. (1959). Hydrodynamic aspects of boiling heat transfer. *AECU-4439.*

16

FLOW BOILING: ONSET OF NUCLEATION AND HEAT TRANSFER

16.1 Introduction

Flow boiling typically takes place in a vertical boiler tube. The appearance of the boiling mixture for this case is shown in Fig. 16.1. Referring to Figure 16.1, the distinct areas of the flow are (see Collier, 1981):

1. single-phase liquid,
2. bubbly flow,
3. plug flow,
4. churn flow,
5. annular flow,
6. dispersed-drop flow, and
7. single-phase vapour.

The important points in the flow are, again referring to Fig. 16.1, as follows.

(*a*) Bubble nucleation begins when the thermodynamic quality x_T is less than zero. The thermodynamic quality is given by

$$x_T = \frac{h - h_{1,sat}}{\lambda},\qquad(16.1)$$

where h is the enthalpy of the flowing mixture (J/kg), and $h_{1,sat}$ is the saturation enthalpy of the liquid phase (J/kg). Thus the nucleation process starts when the liquid in the bulk flow (away from the walls) is still subcooled.

(*b*) $x_T = 0$.

(*c*) Bubble nucleation ceases and the boiling process becomes convective in character.

(*d*) The liquid film dries out. This is the dryout, burnout, or critical heat flux condition which is discussed separately in Chapter 17.

(*e*) $x_T = 1$.

(*f*) The last liquid drops evaporate.

The facts that bubble nucleation starts when $x_T < 0$ and liquid drops persist when $x_T > 1$ demonstrate that there is thermodynamic non-equilibrium in these areas: both the liquid and vapour are not saturated

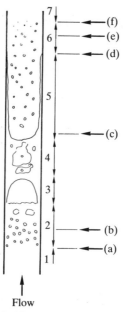

Fig. 16.1. Flow boiling in a vertical tube.

and so not in equilibrium with each other. The temperature variation of the tube wall and the fluid up the tube can be plotted as in Fig. 16.2.

From Fig. 16.2 it can be seen that there is a large increase in wall temperature when dryout occurs: this increase may be many hundreds of degrees. Also, before dryout occurs the difference between wall temperature and the saturation temperature is gradually decreasing. Therefore, the boiling heat transfer coefficient is increasing; this is a conse-

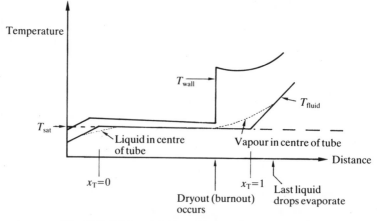

Fig. 16.2. Wall and fluid temperature in flow boiling.

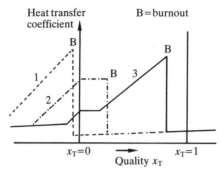

Fig. 16.3. Variation of heat transfer coefficient with quality in flow boiling.

quence of the liquid film in annular flow becoming thinner as the quality increases. One convenient way of expressing the heat transfer coefficient variation along the tube is to plot it against the thermodynamic quality, as in Fig. 16.3. In this figure, lines 1, 2, and 3 represent different heat fluxes. At the highest heat flux, curve 1, burnout occurs before x_T is equal to zero. At a lower heat flux, curve 2, burnout occurs after $x_T = 0$. In the region of positive quality nucleate boiling occurs, and this ceases at burnout. At the lowest heat flux, curve 3, there are five distinct regions. These are (starting from $x_T < 0$):

(a) single-phase liquid forced convection (in which the heat transfer coefficient is almost constant);

(b) subcooled boiling, in which the heat transfer coefficient increases as the bulk fluid approaches the saturation temperature;

(c) saturated nucleate boiling in which the heat transfer coefficient is almost constant.

(d) saturated convective boiling in which the heat transfer coefficient increases slowly; and

(e) post-burnout heat transfer, in which the heat transfer coefficient is low. This regime gradually merges into a single-phase vapour forced-convection regime.

16.2 Onset of nucleate boiling

As in pool boiling (see § 14.7) the wall temperature must be some way above the saturation temperature before bubble growth can occur. In flow boiling the situation is complicated by the fact that there is a temperature profile in the liquid near the wall (see Fig. 16.4). There is also, in general, a large range of cavity sizes available. As the bubbles grow from these cavities they protrude into the cooler liquid away from the wall. If we restrict our attention to the thin layer next to the wall where heat conduction through the liquid is dominated by molecular

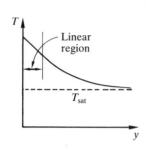

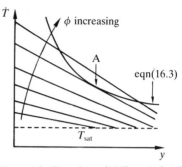

Fig. 16.4. Temperature variation with distance from the heated wall during flow.

Fig. 16.5. Equations (16.2) and (16.3) for temperature plotted against distance from the wall.

thermal conductivity rather than by turbulent convection, then the temperature T near the wall is given by

$$T = T_w - \frac{\phi y}{k_1},$$
(16.2)

where T_w is the wall temperature (K), ϕ is the heat flux (W/m^2) from the wall into the fluid, y is the distance from the wall (m), and k_1 is the liquid thermal conductivity (W/m K). From § 14.7, the condition for the bubble of radius y to grow is that

$$T > T_{sat} + \frac{2\sigma}{y} \frac{T_{sat}}{\lambda \rho_g}$$
(16.3)

Equations (16.2) and (16.3) can be plotted together as in Fig. 16.5. Here, the straight lines are eqn (16.2) plotted at a number of heat fluxes, and the curved line is eqn (16.3).

One suggestion is that boiling can begin when the curves touch tangentially as at A in Fig. 16.5. At this position the liquid at the outer point of the hemispherical bubble is just hot enough for the bubble to continue growing. At A, eqns (16.2) and (16.3) give the same temperature, so

$$T_w - \frac{\phi y}{k_L} = T_{sat} + \frac{2\sigma}{y} \frac{T_{sat}}{\lambda \rho_g}$$
(16.4)

or, rearranging,

$$T_w - T_{sat} = \frac{1}{y} \left(\frac{2\sigma T_{sat}}{\lambda \rho} + \frac{\phi}{k_L} y^2 \right).$$
(16.5)

Also, the temperature gradients are equal, so

$$-\frac{\phi}{k_L} = -\frac{2\sigma \, T_{sat}}{y^2 \, \lambda \rho_g},$$ (16.6)

or

$$y^2 = \frac{2\sigma T_{sat} k_1}{\phi \lambda \rho_g}.$$ (16.7)

Substituting for y from eqn (16.7) into eqn (16.5) gives

$$T_w - T_{sat} = \left(\frac{\phi \lambda \rho_g}{2\sigma T_{sat} k_1}\right)^{\frac{1}{2}} \left(\frac{2\sigma T_{sat}}{\lambda \rho_g} + \frac{2\sigma T_{sat}}{\lambda \rho_g}\right),$$ (16.8)

or

$$(T_w - T_{sat})^2 = \frac{8\sigma T_{sat}\phi}{\lambda \rho_g k_1}.$$ (16.9)

Thus nucleate boiling will occur if

$$\Delta T_{sat} > [(8\sigma T_{sat}\phi)/(\lambda \rho_g k_1)]^{\frac{1}{2}}.$$ (16.10)

This equation was first derived by Davis and Anderson (1966).

Example

Calculate ΔT_{sat} for water at 1 bar, for $\phi = 10^5 \, \text{W/m}^2$. The physical properties required are:

$\sigma = 0.059 \, \text{N/m}$,
$T_{sat} = 373 \, \text{K}$,
$\lambda = 2.256 \times 10^6 \, \text{J/kg}$,
$\rho_g = 0.598 \, \text{m}^3/\text{kg}$, and
$k_1 = 0.681 \, \text{W/m K}$.

From eqn (16.10)

$$\Delta T_{sat} = \left(\frac{8\sigma T_{sat}\phi}{\lambda \rho_g k_1}\right)^{\frac{1}{2}} = 4.38 \, \text{K},$$

and the cavity radius operating is given by eqn (16.7)

$$R = \left(\frac{2\sigma T_{sat} k_1}{\phi \lambda \rho_g}\right)^{\frac{1}{2}} = 15 \, \mu\text{m}.$$

In pool boiling (§ 14.7), it was found that for water at 1 bar, if it was assumed that $\Delta T_{sat} = 5 \, \text{K}$, then $R = 6.5 \, \mu\text{m}$. In the current situation the cavity size is much greater because of the effect of the temperature gradient in the liquid (see Fig. 16.6). At the outer edge of the bubble the

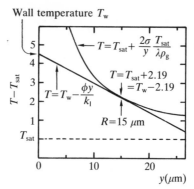

Fig. 16.6. Temperature profiles for the example in § 16.2.

temperature is half-way between the wall temperature and the saturation temperature. A further example is given in Example 8 in Chapter 25.

This nucleation theory, first fully developed by Davis and Anderson (1966), will work if the following are true.

(a) *There is a full range of cavity sizes.* It does not work, for example, in liquid metals which are very well-wetting and thus very efficient at expelling air from the cavities.

(b) *The temperature distribution in the liquid is indeed linear.* Eventually, of course, the temperature distribution joins evenly with the bulk temperature as shown in Fig. 16.4.

16.3 Nucleate and convective boiling

It is commonly, but not universally, believed that there are, as assumed up to now, two types of behaviour in flow boiling.

(a) *Nucleate boiling—in which bubbles are formed by nucleation at the solid surface.* In highly subcooled boiling these bubbles rapidly collapse, transferring their latent heat to the liquid phase and thus heating it up towards the saturation temperature.

(b) *Convective boiling—in which heat is transferred by conduction and convection through a thin liquid film.* Evaporation then takes place at the liquid–vapour interface.

The conditions under which these types of boiling occur will be discussed further in Chapter 19. There, however, and in § 16.1, it is assumed that only one of the boiling types occurs at once, and that at some point the mechanism suddenly switches from one type of boiling to the other. In fact, the mechanisms can coexist and, as the quality increases, convective boiling gradually supplants nucleate boiling.

16.4 Calculation of flow-boiling heat transfer coefficients

It is now clear that a good procedure for calculating flow-boiling coefficients must have some elements of a nucleate-boiling calculation and some elements of a flow-boiling calculation. This was provided by Chen (1963), whose method (or correlation) is probably the best available. He suggested that the two mechanisms work in parallel, and so

$$h_B = h_{NB} + h_{FC}, \qquad (16.11)$$

where h_B is the total boiling heat transfer coefficient, h_{NB} is the nucleate-boiling heat transfer coefficient, and h_{FC} is the forced-convection heat transfer coefficient.

$$h_{NB} = S h_{FZ}, \qquad (16.12)$$

where S is the suppression factor, and h_{FZ} is the nucleate-boiling heat transfer coefficient calculated from the Forster–Zuber equation. The suppression factor gradually decreases from 1 to 0 as the quality increases; the detailed definition is given later. The equation for h_{FZ} is

$$h_{FZ} = \frac{0.001\,22\,\Delta T_{sat}^{0.24}\Delta p_{sat}^{0.75} C_{pl}^{0.45} \rho_l^{0.49} k_l^{0.79}}{\sigma^{0.5}\lambda^{0.24}\mu_l^{0.29}\rho_g^{0.24}}. \qquad (16.13)$$

This equation is dimensionally consistent, and so if SI units are used on the right hand side the units of h_{FZ} will be W/m^2 K. Note that C_{pl} is the liquid specific heat, k_l is the liquid thermal conductivity, $\Delta T_{sat} = T_w - T_{sat}$, and Δp_{sat} is the difference in saturation pressure corresponding to ΔT_{sat}. Δp_{sat} is shown on the vapour pressure curve (see Fig. 16.7).

Apparently, eqn (16.13) does not agree with the general findings (described in § 14.8) about nucleate-boiling heat transfer coefficients that

$$h \propto \Delta T_{sat}^{a-1}, \qquad (16.14)$$

where $a = 3$ or 3.33. However, it must be recognized that eqn (16.13) also contains Δp_{sat}, which is a strong function of ΔT_{sat}. For the forced

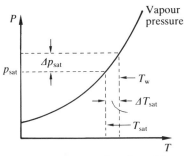

Fig. 16.7. Vapour pressure curve; explanation of the meaning of Δp_{sat}.

convection part of the total heat transfer coefficient,

$$h_{FC} = h_1 F, \tag{16.15}$$

where h_1 is the single-phase liquid convective heat transfer coefficient based on the mass flow rate of liquid in the two-phase flow (not the total two-phase flow rate). This can be calculated from, for example, the Dittus–Boelter equation

$$Nu_1 = 0.023 Re_1^{0.8} Pr_1^{0.4}, \tag{16.16}$$

where

$$Nu_1 = \frac{h_1 d}{k_1}; \tag{16.17}$$

$$Re_1 = \frac{G(1-x)d}{\mu_1}; \tag{16.18}$$

$$Pr_1 = \frac{\mu_1 C_{pl}}{k_1}; \tag{16.19}$$

and F is the two-phase heat-transfer coefficient multiplier, which is greater than unity. Chen's special contribution was to suggest ways of calculating the factors S and F. F, he suggested was a function of the Martinelli parameter X

$$X^2 = \frac{(dp/dz)_1}{(dp/dz)_g}, \tag{16.20}$$

where the two pressure gradients were calculated assuming turbulent flow. $(dp/dz)_1$ is the frictional pressure gradient in single-phase liquid flow, where the flow rate is equal to the liquid flow rate in the two-phase flow (not the total two-phase flow rate). $(dp/dz)_g$ is defined similarly for the gas. If the friction factor for each phase is proportional to $Re^{-0.2}$, then X can be calculated simply from

$$X = \left(\frac{1-x}{x}\right)^{0.9} \left(\frac{\rho_g}{\rho_l}\right)^{0.5} \left(\frac{\mu_l}{\mu_g}\right)^{0.1}. \tag{16.21}$$

The variation of F with X is shown in Fig. 16.8. Thus if the quality and the physical properties are known, then F can be found.

In order to find the suppression factor S, the liquid Reynolds number Re_1 is calculated

$$Re_1 = \frac{G(1-x)d}{\mu_1}. \tag{16.22}$$

A two-phase Reynolds number Re_{TP} is then given by

$$Re_{TP} = Re_1 F^{1.25}, \tag{16.23}$$

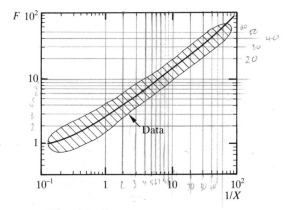

Fig. 16.8. Chen correlation for F.

where F is the heat-transfer multiplier previously found from Fig. 16.8. S is then a function of Re_{TP} (as shown in Fig. 16.9). This then enables the total boiling heat transfer coefficient h_B to be calculated.

As x increases X decreases,

$\quad$ and $\dfrac{1}{X}$ increases,

$\quad$ therefore F increases,

$\quad\quad$ and Re_{TP} increases,

$\quad$ therefore S decreases.

Thus, as the quality goes up, the relative contribution of the nucleate boiling mechanism goes down. The Chen method is a complicated one; a detailed illustration of its use is given in Example 8 in Chapter 25.

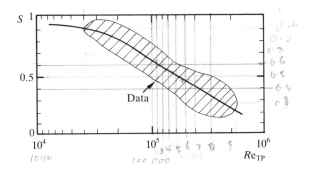

Fig. 16.9. Chen correlation for S.

16.5   Application to subcooled boiling

The Chen method is for saturated boiling. For subcooled boiling, a suitable procedure for calculating the heat flux ϕ seems to be to assume that

$$\phi = h_{\mathrm{l}}(T_{\mathrm{w}} - T_{\mathrm{B}}) + h_{\mathrm{NB}}(T_{\mathrm{w}} - T_{\mathrm{sat}}), \qquad (16.24)$$

where T_{w} is the wall temperature and T_{B} is the bulk liquid temperature, and to calculate S on the basis that

$$Re_{\mathrm{TP}} = Re_{\mathrm{l}}.$$

This procedure implies that the whole of the actual temperature difference $(T_{\mathrm{w}} - T_{\mathrm{B}})$ goes into the convective heat flux, but only part of it into the nucleate component. The heat flux gradually and smoothly reaches the saturated value as the bulk temperature reaches the saturation temperature.

References

CHEN, J. C. (1963). A correlation for boiling heat transfer to saturated fluids in convective flow. *ASME paper 63-HT-34.* Presented at 6th National Heat Transfer Conference, Boston.

COLLIER, J. G. (1981). *Convective boiling and condensation,* (2nd edn). McGraw-Hill, New York.

DAVIS, E. J. and ANDERSON, G. H. (1966). The incipience of nucleate boiling in forced convection flow. *AIChE J.* **12,** 774–80.

17

CRITICAL HEAT FLUX IN FLOW BOILING

17.1 Introduction

In Chapter 16 (see, for example, Fig. 16.2) it was seen that the heat transfer coefficient in flow boiling could fall rapidly and the wall temperature increase rapidly at some point along the heated channel. This phenomenon is known by many names, none of them entirely satisfactory.

(a) *Burnout*. This, however, implies that the physical surface is destroyed; this does not always occur.

(b) *Dryout*. This, however, implies a particular mechanism which does not always occur.

(c) *CHF—critical heat flux*. This is a cumbersome term but it is widely used to refer to the phenomenon and not just the value of the heat flux at which it occurs.

(d) *DNB—departure from nucleate boiling*. This again implies a particular mechanism which does not always occur.

(e) *Boiling crisis*. This is widely used in the USSR and is a very descriptive term. However, it is commonly used in the form *boiling crisis of the first* (or *second*) *kind*, and it is then often not clear what is meant.

CHF seems to have two distinctive characters.

(a) At low quality, when it is associated with subcooled boiling or saturated nucleate boiling, it has strong similarities to pool-boiling critical heat flux, both in mechanism (formation of bubbles at the heated surface impeding liquid flow to the surface) and in behaviour (there is a hysteresis effect as the heat flux is increased and then decreased (see Fig. 17.1(a)).

(b) At high quality, it is associated with convective boiling in annular flow in which the liquid film dries out and there is no hysteresis (see Fig. 17.1(b)).

17.2 Mechanisms in the subcooled region

When critical heat flux occurs in subcooled (or low quality) flow there are certain resemblances to pool-boiling critical heat flux, as already noted.

155

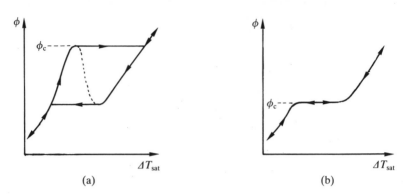

Fig. 17.1. Boiling curves for nucleate and convective boiling.

A number of detailed mechanisms for this type of critical heat flux have been suggested (see Hewitt 1978).

(*a*) *Near-wall bubble crowding and vapour blanketing.* A kind of bubble boundary layer builds up at the heated wall. This becomes so thick and dense that it effectively stops fresh liquid reaching the surface. However, how this occurs is still uncertain, and the full details of the mechanism are unclear. Is there, for example, some form of separation occurring in the 'boundary layer' which leads to stagnation?

(*b*) *Overheating at a nucleation site.* It has been suggested that at very high heat fluxes, the nucleation site, which is locally dry, heats up so much during the bubble growth phase that it cannot be rewetted properly when the bubble has departed. The subject of rewetting hot surfaces is discussed further in Chapter 21.

(*c*) *Vapour clot or slug formation.* In horizontal flow, particularly, a vapour clot or slug near a heated wall (the top wall in horizontal flow) can approach the wall very closely, so that the intermediate liquid film can evaporate; the wall is then too hot to rewet.

It seems probable that all of these mechanisms occur: the first being the usual one, the second occurring at very high heat fluxes, and the third at low flow rates (where the vapour is not swept rapidly along) or in horizontal flow.

17.3 Mechanisms in the high quality regions

Observations of transparent test sections and flow pattern maps show that, for most critical heat flux cases where the quality is greater than 10 %, the flow pattern is annular. Many detailed studies have suggested that the critical heat flux occurs when the combined effects of entrain-

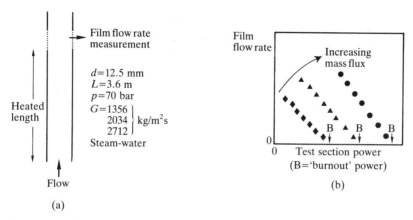

Fig. 17.2. Film flow rates near high quality critical heat flux.

ment, deposition, and the evaporation of the film make the film flow rate go gradually and smoothly to zero. The evidence for this is threefold.

(*a*) *Measurements of liquid film flow rate* (Hewitt and Hall Taylor 1970). Experiments were conducted in a vertical heated tube in which the liquid film flow rate could be measured at the top of the heated section. This measurement was made by sucking the film off through a porous section of tube wall, as described for adiabatic flow in § 9.5 (see Fig. 17.2(a)). The results are shown schematically in Fig. 17.2(b). The critical heat flux point, where the heat transfer coefficient drops dramatically, occurs at the test section power which gives zero film flow rate at the outlet of the tube.

(*b*) *'Cold patch' experiments* (Bennett *et al.* 1967). Experiments were again performed in a vertical tube with steam–water flow. The tube diameter here was 9.3 mm and the pressure was 3.4 bar. Four separate test sections were used, as shown in Fig. 17.3. The results for the critical heat flux, here defined as the critical power divided by the heated area, are shown in Table 17.1.

A, C, and D all have a heated length of 1.8 m and so are directly comparable, but the position of the cold patch can apparently lead to an increase or a decrease in the critical heat flux. The explanation lies with the details of the annular flow and the variation of the entrained liquid flow with the tube length. Figure 17.4 shows the experimentally determined variation of entrained liquid flow with quality in the four test sections, and also the equilibrium entrained liquid flow rate. CHF will occur when the line for each tube meets the total liquid line, because then all the liquid is entrained and the film flow is zero.

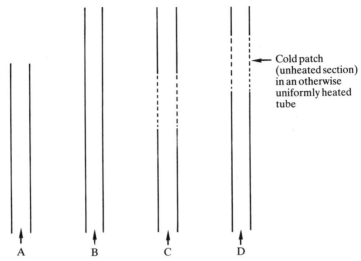

Cold patch
(unheated section)
in an otherwise
uniformly heated
tube

Fig. 17.3. Test sections for cold patch experiment.

TABLE 17.1. *Results of cold patch experiments*

Test section	A	B	C	D
Tube length (m)	1.8	2.4	2.4	2.4
Cold patch (unheated section) length (m)	—	—	0.6	0.6
Distance of cold from tube inlet (m)	—	—	1.1	1.5
Critical heat fliux ϕ_c (kW/m²)	652	400	617	690

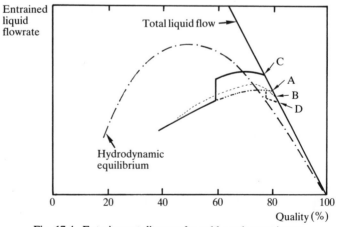

Fig. 17.4. Entrainment diagram for cold patch experiment.

Heating the flow increases the quality and the point representing the flow conditions moves along the curve to the right. The system also tries to reach the hydrodynamic equilibrium line. Thus when the operating line crosses the equilibrium line it does so with zero gradient. Over the length of a cold patch the quality does not change, so the point can move vertically—again towards the equilibrium line. Test section C has the cold patch near the entrance, so it is encountered at low quality. The liquid entrainment rises over the length of this cold patch and the critical heat flux is lower than for a uniformly heated tube with the same heated length (test section A). Test section D, on the other hand, has the cold patch nearer the exit, so it is encountered at higher quality. The entrainment falls over the length of this cold patch and the critical heat flux is higher than for the uniformly heated tube. This is the reason why the cold patch can increase or decrease the critical heat flux.

(c) *Physically based models of the flow*. The third piece of evidence for the film dryout mechanism is the relative success of models based on it for predicting actual values of critical heat flux, see Chapter 19.

17.4 Parametric trends and forms of correlation (see Collier 1981)

The main parametric trends for the critical heat flux for a uniformly heated tube are illustrated in Fig. 17.5.

(a) *Figure 17.5(a)*. The critical heat flux usually varies linearly with the inlet subcooling Δh_s (J/kg). The inlet subcooling is the liquid saturation enthalpy minus the actual liquid enthalpy at the test section entrance.

(b) *Figure 17.5 (b and c)*. The critical heat flux increases with mass flux and tube diameter.

(c) *Figure 17.5(d)*. The critical heat flux asymptotes to zero as the tube length increases. The critical power P_c, which is given by

$$P_c = \pi d L \phi_c, \qquad (17.1)$$

asymptotes to a power sufficient to vaporize all the liquid and thus produce a quality of unity.

(d) *Figure 17.5(e)*. For steam–water flows the critical heat flux goes through a maximum at about 70 bar. From § 15.7 it can be seen that the pool-boiling critical heat flux for steam–water also passes through a maximum at about 70 bar. The connection between these facts is not clear.

The common forms of correlation for critical heat flux in flow boiling can be expressed as two graphical plots, as in Fig. 17.6. In this figure x_c is quality at which the critical condition occurs, and L_B is the length over which boiling occurs (m).

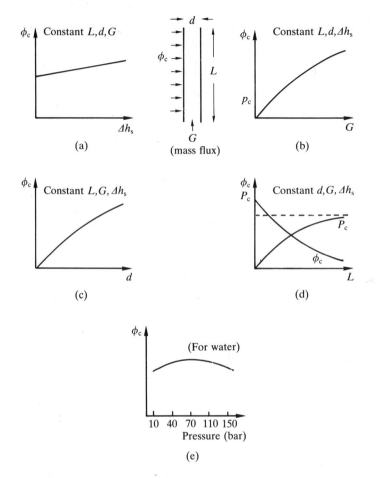

Fig. 17.5. Parametric trends in the critical heat flux.

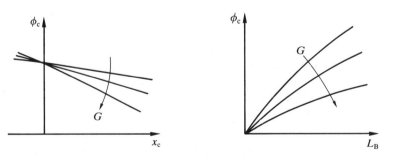

Fig. 17.6. Forms of critical heat flux correlation.

Figure 17.6(a) suggests that CHF is a purely local condition: if the local values of x and ϕ correspond to the critical values, then CHF will occur.

Figure 17.6(b) suggests that CHF is an integral process: it contains the boiling length which encompasses some information about the history of the flow.

Most experiments are, however, performed in uniformly heated tubes. A heat balance over the boiling length L_B gives

$$\pi d L_B \phi_c = \frac{\pi d^2}{4} G \lambda x_c \qquad (17.2)$$

or

$$L_B \phi_c = \frac{dG\lambda}{4} x_c, \qquad (17.3)$$

and so a correlation of ϕ_c against x_c contains no less information than one of ϕ_c against L_B. The most usual form of correlation encountered in practice is one of ϕ_c against x_c, as in Fig. 17.6(a). This type of correlation is known as a 'local conditions' cocrrelation. The critical heat flux is assumed to be a function of the local conditions only.

17.5 Local conditions correlations

A typical correlation of this type is that due to MacBeth (1963). A simplified version is

$$\phi_c = A\lambda G^{\frac{1}{2}}(1-x). \qquad (17.4)$$

This was developed for water, and the constant A is approximately $0.25 \, \text{kg}^{\frac{1}{2}}/\text{m} \, \text{s}^{\frac{1}{2}}$. On its own it is of little use as it relates two unknown quantities ϕ_c and x. It can be combined with a heat balance to eliminate ϕ_c or the quality x. The quality is usually eliminated. It can, however, be used to decide where in the tube the critical condition will occur. Both uniformly heated tubes and non-uniformly heated tubes are considered.

(a) *Uniformly heated tubes* (see Fig. 17.7). Figure 17.7(a) shows the basic local conditions correlation (eqn (17.4)). Figure 17.7(b) shows the variation of quality with length for three different heat fluxes. The result of eliminating the quality is shown in Fig. 17.7(c). The critical heat flux can be compared with the actual heat flux shown in Fig. 17.7(d). The comparison is shown directly in Fig. 17.7(e). At heat flux levels 1 and 2 the two lines do not touch or intersect, so CHF does not occur. Raising the heat flux to level 3 causes CHF to occur for the first time at the exit of the tube. This is a general result: in uniformly heated vertical tubes, CHF always occurs first at the tube outlet.

(b) *Non-uniform heating.* In this case a sinusoidal variation of heat flux along the tube is assumed. The result of repeating the previous steps

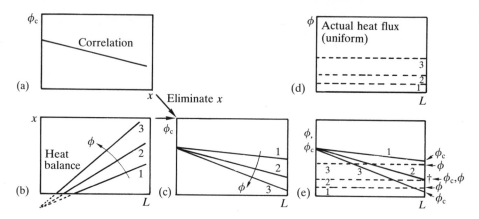

Fig. 17.7. Position of the CHF for a uniformly heated tube. † CHF first occurs at the end of the tube.

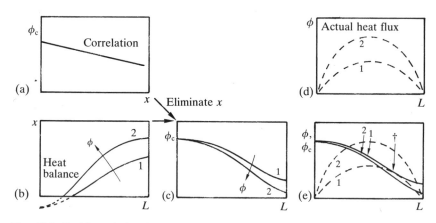

Fig. 17.8. Position of the CHF for a non-uniformly heated tube. † CHF first occurs upstream of the end of the tube.

is shown in Fig. 17.8. In Fig. 17.8(e) at the least flux level, level 1, the actual heat flux and the critical heat flux just touch, so CHF occurs. It occurs some distance upstream from the end of the tube. As the heat flux is increased to level 2 the CHF area spreads upstream and downstream. Whenever the heat flux falls with distance along the test section, there is the possibility that CHF may first occur at a point before the tube outlet.

17.6   Russian CHF results (see Hewitt 1978)

Russian and other workers have reported CHF results which, when plotted as ϕ_c against x, have the form shown in Fig. 17.9(a). The most

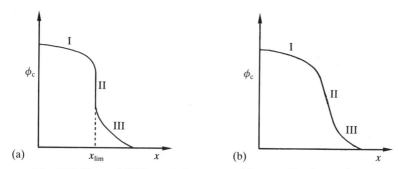

Fig. 17.9. Form of CHF correlation proposed by some Russian workers.

striking feature is the existence of a vertical portion to the curve (region II) at a 'limiting quality'. Many of the data exhibiting such a limiting quality were obtained in experiments for which the fluid at the tube inlet was not liquid water as is usual, but was a mixture of steam and water. Such 'mixed' inlet conditions are used for convenience—to reduce the power requirements of the experiments and to enable shorter test sections to be used. Unfortunately, the method of mixing the water and the steam at inlet affects the initial fraction of liquid entrained, and this affects the critical heat flux results.

It seems probable that the shape of the curve is actually as shown in Fig. 17.9(b). There is a steep gradient at II, but no limiting quality. Experimental scatter makes the position difficult to resolve. One of the greatest difficulties with the limiting quality hypothesis is to think up a plausible mechanism that might account for the apparent sudden instability of the liquid film once a certain quality has been reached.

17.7 Calculation of critical heat flux

Due to the vast amount of data collected for steam–water flow in vertical tubes, it is possible to recommend calculation methods for this situation.

(a) *Bowring* (1972) *correlation.* This has the form

$$\phi_c = \frac{A + B\Delta h_s}{C + L},\tag{17.5}$$

where A, B, and C are functions of p, λ, d, and G. Note that the Bowring correlation can be rearranged into the local conditions form, where the critical heat flux is a function of the local quality. The details of the Bowring correlation are given in Appendix C, and in Example 8 in Chapter 25.

(b) *Groeneveld* (1982) *tabular method.* Groeneveld has developed a method which gives tables for the function

$$\phi_c = f(p, G, x). \tag{17.6}$$

There are correction factors for tube diameter d and tube length L. Note that this, again, is basically a kind of local condition form of expression. Groeneveld's tabular method is a development of earlier tables published by Russian workers (see Doroshchuk *et al.* 1975; Collier 1981).

For fluids other than steam–water, there are two options available.

(a) *Ahmad* (1973) *scaling rules.* Of the very many possible dimensionless groups for characterizing critical heat flux, Ahmad identified the important ones as

$$\frac{L}{d}, \quad \frac{\rho_1}{\rho_g}, \frac{\Delta h_s}{\lambda}, \quad \psi, \quad \text{and} \quad \frac{\phi_c}{G\lambda}.$$

Here

$$\psi = \frac{Gd}{\mu_1} \left(\frac{\mu_1^2}{\sigma d \rho_1} \right)^{\frac{2}{3}} \left(\frac{\mu_g}{\mu_1} \right)^{\frac{1}{5}}. \tag{17.7}$$

If the first four dimensionless groups are the same in the two different fluid systems, then the fifth dimensionless group will also be the same in the two systems. Hence to calculate ϕ_c for a fluid X from information about critical heat flux in water the following procedure is recommended.

(i) Make L/d the same for the water system and the fluid X system. Normally, the two systems have equal lengths and equal diameters.

(ii) Make ρ_1/ρ_g the same for the water system and the fluid X system. This determines the water pressure.

(iii) Make $\Delta h_s/\lambda$ the same for the water system and the fluid X system. This determines the water inlet enthalpy.

(iv) make ψ the same for the water system and the fluid X system. This determines the water mass flux.

(v) Calculate ϕ_c for the now fully determined water conditions. This can be done by, for example, the Bowring correlation described above.

(vi) Hence find ϕ_c for the fluid X, knowing that

$$\left(\frac{\phi_c}{G\lambda} \right)_{\text{water}} = \left(\frac{\phi_c}{G\lambda} \right)_{\text{X}}. \tag{17.8}$$

(b) *Katto generalized correlation* (see Katto and Ohne 1984). In a long series of papers Katto has assumed that

$$\phi_c = XG(\lambda + K\Delta h_s), \tag{17.9}$$

and then correlated X and K. Note, therefore, that he has assumed that the critical heat flux varies linearly with the inlet subcooling. This is usually a good assumption. He further assumed that X and K were functions of three dimensionless groups

$$X, K = f\left(\frac{L}{d}, \frac{\rho_1}{\rho_g}, \frac{\sigma\rho_1}{G^2L}\right),$$ (17.10)

The basic Katto equation (eqn (17.9)) can be rearranged to give

$$\frac{\phi_c}{G\lambda} = X + XK\frac{\Delta h_s}{\lambda}.$$ (17.11)

It is now evident that the Ahmad scaling rules and the Katto form are very similar. Of Ahmad's five dimensionless groups, Katto uses four. The disagreement is about the fifth:

Ahmad	Katto

$$\frac{Gd}{\mu_1}\left(\frac{\mu_1^2}{\sigma d\rho_1}\right)^{\frac{2}{3}}\left(\frac{\mu_g}{\mu_1}\right)^{\frac{1}{3}} \qquad\qquad \frac{\sigma\rho_1}{G^2L}$$

The Katto correlation has probably been better tested than the Ahmad scaling rules. Katto used test data for:

water,
ammonia,
benzene,
ethanol,
helium,
hydrogen,
nitrogen,
R12, R21, R22, and R113, and
potassium.

He seemed to obtain good results for most data. The Katto correlation is given in detail in Appendix D, and in Example 8 in Chapter 25.

References

AHMAD, S. Y. (1973). Fluid to fluid modelling of critical heat flux, a compensated distortion model. *Int. J. Heat Mass Transfer* **16**, 641–61.
BENNETT, A. W., HEWITT, G. F., KEARSEY, H. A., KEEYS, R. K. F., and PULLING, D. J. (1967). Studies of burnout in boiling heat transfer to water in round tubes with non-uniform heating. *Trans. Instn Chem. Engnrs* **45**, 319–33 (see also AERE-R5076, 1966).
BOWRING, R. W. (1972). A simple but accurate round tube uniform heat flux dryout correlation over the pressure angle 0.7–17 MN/m². *AEEW-R789*.

COLLIER, J. G. (1981). *Convective boiling and condensation* (2nd edn). McGraw-Hill, New York.

DOROSHCHUK, V. E., LEVITAN, L. L., and LANTZMAN, F. P. (1975). Investigations into burnout in uniformly heated tubes. *ASME paper 75-WA/HT-22.*

GROENEVELD, D. C. (1982). A general CHF prediction method for water suitable for reactor accident analysis. *CENG Report DRE/STT/SETRE/82-2-E/DG.*

HEWITT, G. F. (1978). Critical heat flux in flow boiling. *Sixth Int. Heat Transfer Conference, Toronto* **6**, 143–71.

HEWITT, G. F. and HALL TAYLOR, N. S. (1970). *Annular two-phase flow.* Pergamon Press, Oxford.

KATTO, Y. and OHNE, H. (1984). An improved version of the generalised correlation of critical heat flux for convective boiling in uniformly heated vertical tubes. *Int. J. Heat Mass Transfer* **27**, 1641–8.

MACBETH, R. V. (1963). Burnout analysis, III The low velocity burnout regime. *AEEW-R222.*

18

MIXTURE EFFECTS IN BOILING

18.1 Introduction

The boiling of a mixture of liquids can exhibit significant differences from boiling of a single, pure component. For example, the pool-boiling curves for a mixture and for a pure liquid with the same physical properties ('the equivalent pure fluid') are shown in Fig. 18.1. The following points can be noted.

(a) For a mixture, the onset of nucleate boiling occurs at a larger temperature difference than for the pure fluid.

(b) The pool-boiling heat transfer coefficient in a mixture is lower than for the pure fluid.

(c) The critical heat flux is affected by the mixing effects, but may be increased or decreased.

The differences are related to the phase equilibrium effects. This can be expressed on a temperature–composition diagram.

(a) For a reasonably ideal mixture the temperature–composition diagram is shown in Fig. 18.2. The liquid and vapour compositions (x and y) are related as shown. When a liquid of composition x boils, the vapour in equilibrium with the liquid has a composition y.

(b) More complicated mixtures commonly occur which have an azeotrope. This is a liquid which has a composition such that the liquid and the vapour in equilibrium with it, have the same composition. Azeotropes are useful because they behave in many ways like pure fluids because there is no composition difference between the phases. Examples of systems having azeotropes are ethanol–water, ethanol–benzene, and acetone–chloroform. Figure 18.3 shows the temperature–composition diagram for the ethanol–benzene system.

18.2 Onset of nucleate boiling

In § 14.7 it was found that, at the onset of nucleate boiling, boiling will occur if the temperature difference $\Delta T_{\text{sat}}(= T_{\text{w}} - T_{\text{sat}})$ exceeds $(\mathrm{d}T/\mathrm{d}p)(2\sigma/R)$. If, as is found at the onset of nucleate boiling,

$$(\Delta T_{\text{sat}})_{\text{mixture}} > (\Delta T_{\text{sat}})_{\text{pure liquid}}, \tag{18.1}$$

167

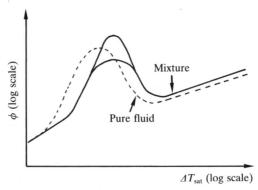

Fig. 18.1. Boiling curves for a liquid mixture and for a pure liquid.

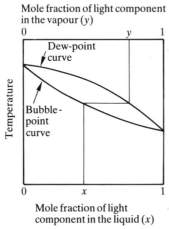

Fig. 18.2. Phase equilibrium diagram for a normal mixture.

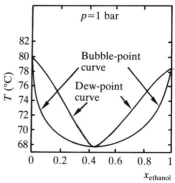

Fig. 18.3. Phase equilibrium diagram for a mixture which has an azeotrope.

then the difference must be caused by one of the parameters: dT/dp, σ, or R.

(a) dT/dp (*the slope of the saturation line*). Actually, this is not significantly changed by the mixture effects.

(b) σ (*surface tension*). This may alter significantly, particularly when a small amount of organic liquid is added to water, but in many systems (such as ethanol–benzene) the change in surface tension is small.

(c) R (*cavity radius*). Of course, the physical size of the cavities in the surface is fixed, but for the cavities to operate as nucleation sites they must contain trapped gas (usually air). Whether air will be trapped depends on the cavity geometry (the size and shape), the surface tension, and on the contact angle ϕ (see Fig. 18.4).

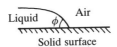

Fig. 18.4. An azeotrope contact angle.

As the contact angle $\phi \to 0$, the liquid 'wets' the surface more efficiently and displaces the air from the cavities. It is generally thought that the change in contact angle, which alters the effective value of the cavity radius, is the important change which leads to an increased value of the wall superheat at the onset of nucleate boiling (see Thome and Shock 1984).

18.3   Nucleate-boiling heat transfer coefficients

Typical results for the ethanol–benzene system are shown in Fig. 18.5 (see Hui 1983). Note that ethanol–benzene is a system with an azeotrope. At the azeotropic composition, the heat transfer coefficient is very close to the line joining the pure fluids. Hence, the azeotrope is behaving like the equivalent pure fluid.

The most common method used for calculating nucleate boiling coefficients for mixtures is due to Stephan and Korner (1969).

$$(\Delta T_{\text{sat}})_{\text{mixture}} = \Delta T_{\text{sat,e}}(1 + \theta), \tag{18.2}$$

where $(\Delta T_{\text{sat}})_{\text{mixture}}$ is the temperature difference (K) for the mixture at a given heat flux, and $\Delta T_{\text{sat,e}}$ is the temperature difference (K) for the equivalent pure fluid at the same heat flux. The equivalent pure fluid is the liquid with the same physical properties as the mixture. $\Delta T_{\text{sat,e}}$ is given by

$$\Delta T_{\text{sat,e}} = x_1 \Delta T_{\text{sat,1}} + x_2 \Delta T_{\text{sat,2}}, \tag{18.3}$$

where x_1 and x_2 are the mole fractions (in the liquid phase) of components 1 and 2, and $\Delta T_{\text{sat,1}}$ and $\Delta T_{\text{sat,2}}$ are the corresponding temperature differences. In eqn (18.2) the dimensionless parameter θ is given by

$$\theta = A \left| y - x \right|, \tag{18.4}$$

where x is the mole fraction of the light component in the liquid phase, and y is the mole fraction of the light component in the vapour phase. The modulus sign ensures that θ is always positive. Also,

$$A = A_0(0.88 + 0.12p), \tag{18.5}$$

where p is the system pressure (bar), and A_0 is an empirical constant depending on the binary mixture. For example, for ethanol–benzene $A_0 = 0.42$. Values for other mixtures are given by Thome and Shock (1984), Collier (1981), and Stephan and Korner (1969).

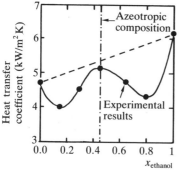

Fig. 18.5. Boiling heat transfer coefficients in ethanol–benzene mixtures.

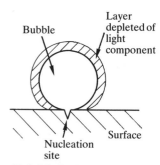

Fig. 18.6. Mechanism of the reduction in heat transfer to a bubble in a mixture.

The mechanism causing reduction in heat transfer coefficients is illustrated in Fig. 18.6. The shaded area around the growing vapour bubble represents a layer of liquid around the bubble which has been partially depleted of the light component, hence the saturation temperature of the liquid in this layer has been increased. The effective temperature difference available, which is $T_w - T_{sat}$, has therefore been reduced and the heat transfer rate falls. The gross effect is that the heat transfer coefficient has been reduced. Of course, the concentration gradient between the main fluid and the depleted layer next to the bubble means that diffusion will occur, and the concentration gradient will be limited by this diffusion. This process is obviously a very complicated one. Thome and Shock (1984) have reviewed some of the complex theories which attempt to model the situation.

18.4 Critical heat flux in pool boiling

Most of the evidence points to the conclusion that pool-boiling critical heat flux is increased by mixture effects, but such findings are not universal. Figure 18.7 shows some butanol–water data from Bobrovich *et al.* (1962). It seems unlikely that wires and flat plates behave very differently, or that the small pressure difference between the two experiments is significant. It can be noted that in Fig. 18.7 the data showing an increase in critical heat flux are more definite and consistent than the data showing a decrease. More representative data are that of Afgan (1968) on the ethanol–benzene system, which has an azeotrope at a mole fraction of ethanol of about 0.45 (see Fig. 18.8). Here, the following can be seen.

(*a*) At the azeotropic composition the critical heat flux is comparatively little changed from the equivalent pure fluid value. Certainly, any possible mixture effect is minimized at the azeotropic composition.

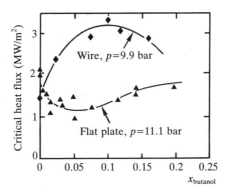

Fig. 18.7. Critical heat flux for butanol–water mixtures.

(b) In the regions away from the azeotrope and the pure fluids, the critical heat flux is increased by the mixture effects.

The mechanisms of the mixture effects on critical heat flux are not well understood; certainly the position is far more complicated than evaluating the mixture properties and substituting them into the single-component critical heat flux equation developed in § 15.4.

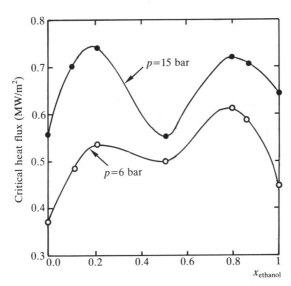

Fig. 18.8. Critical heat flux for ethanol–benzene mixtures.

18.5 Flow boiling

Most process industry flow boiling involves mixtures. In spite of the importance of flow boiling, flow boiling of mixtures has been little

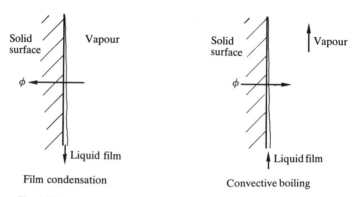

Fig. 18.9. Analogy between film condensation and convective boiling.

studied. It is clear that, like nucleate pool boiling, flow boiling of mixtures leads to lower heat transfer coefficients than are obtained for equivalent pure fluids. General methods for calculating the heat transfer coefficient in mixture boiling have been suggested, but none have been tested against a worthwhile amount of data. For example, from ethylene glycol–water data Bennett and Chen (1980) gave a modification of the single-component Chen correlation. However, the mixture is not very representative as it involves two polar liquids, one of which is relatively non-volatile.

Sardesai *et al.* (1982) have taken up earlier suggestions and recommended that the close analogy between film condensation and convective boiling (see Fig. 18.9) be exploited, and the Bell and Ghaly mixture condensation method be used. This method is explained in detail when condensation is considered in Chapter 22.

References

AFGAN, N. (1968). Boiling heat transfer of binary mixtures. *Int. Summer School on Heat and Mass Transfer in Boundary Layers, Herceg Novi, Yugoslavia, Proceedings,* (ed. N. Afgan, Z. Zaric, and P. Anastasijevic) pp. 761–72, (Vol. 2). Pergamon Press.

BENNETT, D. L. and CHEN, J. C. (1980). Forced convective boiling in vertical tubes for saturated pure components and binary mixtures. *AIChE J.* **26**, 454–61.

BOBROVICH, G. I. *et al.* (1962). PMTF. *Zh. Prikl. Mekh. Tekh. Fiz.* **No. 4,** 108 (quoted by Thorne and Shock 1984).

COLLIER, J. G. (1981). *Convective Boiling and Condensation* (2nd edn). McGraw-Hill.

HUI, T.-O. (1983). M.S. Thesis, Michigan State University (quoted by Thome and Shock 1984).

SARDESAI, R. G., SHOCK, R. A. W., and BUTTERWORTH, D. (1982). Heat and mass transfer in multicomponent condensation and boiling. *Heat Transfer Engineering* **5,** 104–14.

STEPHAN, K. and KORNER, M. (1969). Calculation of heat transfer in evaporating binary liquid mixtures. *Chem. Ing. Tech.* **41,** 409–17.

THOME, J. R. and SHOCK, R. A. W. (1984). Boiling of multicomponent liquid mixtures. *Advances in Heat Transfer* **16,** 59–156.

19

CALCULATION OF BOILING HEAT TRANSFER COEFFICIENTS AND CRITICAL HEAT FLUX USING AN ANNULAR FLOW MODEL

19.1 Introduction

From earlier chapters, it will now be evident that we are in a position to calculate:

(a) boiling heat transfer coefficients in convective boiling using information about the velocity profile in the liquid film and the analogy between heat and momentum transfer; and

(b) critical heat flux by calculating the variation of the liquid film flow rate along the tube due to entrainment, deposition, and evaporation. When the film flow rate is equal to zero, then dryout is assumed to occur (see Hewitt 1978). Here, it is assumed that the critical heat flux is of the high quality type (see § 17.1). It is, of course, not possible to calculate subcooled critical heat flux using an annular flow view of the flow structure.

The boiling heat transfer coefficient is first considered. The initial step is to calculate the eddy diffusivity from the velocity profile, and then to assume that the eddy diffusivity also applies for the transport of heat.

19.2 Eddy diffusivity

In laminar flow, we can write that, for a Newtonian fluid,

$$\tau = \mu \frac{du}{dy}, \tag{19.1}$$

where τ is the shear stress (N/m^2), μ is the molecular viscosity (N s/m^2), and (du/dy) is the velocity gradient (rate of strain) (1/s). When the flow becomes turbulent, the mean shear stress (it now varies with time) is greater than that given by $\mu(du/dy)$, where now (du/dy) is some kind of mean velocity gradient. The apparent viscosity of the fluid has increased. Conventionally, we can write that

$$\tau = (\mu + \rho\varepsilon) \frac{du}{dy}, \tag{19.2}$$

where ε is the eddy diffusivity (m^2/s).

Since, as in Chapter 4, we can define a dimensionless velocity u^+ and a dimensionless distance y^+, where

$$u = \left(\frac{\tau}{\rho}\right)^{\frac{1}{2}} u^+,\qquad(19.3)$$

and

$$y = \frac{\mu}{(\tau\rho)^{\frac{1}{2}}} y^+.\qquad(19.4)$$

Equation (19.2) can be expressed in the form

$$1 = \left(1 + \frac{\rho\varepsilon}{\mu}\right)\frac{du^+}{dy^+}\qquad(19.5)$$

The dimensionless group $(\rho\varepsilon/\mu)$ is sometimes denoted by ε^+, it is a dimensionless eddy diffusivity.

Now, using the three-part universal velocity profile, as in Chapter 4, the values of ε^+ for the various parts of the universal velocity profile can be found.

(a) For $y^+ < 5$, $u^+ = y^+$, $\qquad\qquad$ (19.6)

so $\dfrac{du^+}{dy^+} = 1$ $\qquad\qquad$ (19.7)

and $\varepsilon^+ = 0$. $\qquad\qquad$ (19.8)

(b) For $5 < y^+ < 30$,

$$u^+ = -3.05 + 5 \ln y^+,\qquad(19.9)$$

so $\dfrac{du^+}{dy^+} = \dfrac{5}{y^+}$ $\qquad\qquad$ (19.10)

and $\varepsilon^+ = \dfrac{y^+}{5} - 1.$ $\qquad\qquad$ (19.11)

(c) For $y^+ > 30$,

$$u^+ = 5.5 + 2.5 \ln y^+\qquad(19.12)$$

so $\dfrac{du^+}{dy^+} = \dfrac{2.5}{y^+}$ $\qquad\qquad$ (19.13)

and $\varepsilon^+ = \dfrac{y^+}{2.5} - 1.$ $\qquad\qquad$ (19.14)

The dimensionless eddy diffusivity ε^+ is plotted against the dimensionless distance y^+ from the wall in Fig. 19.1. The disadvantage of the universal velocity profile is clear: the eddy diffusivity profile is not continuous in gradient or even in value. Other, more complicated velocity profiles give

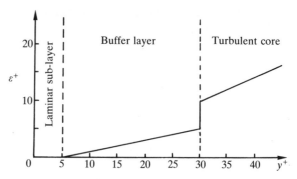

Fig. 19.1. Variation of the dimensionless eddy diffusivity ε^+ with dimensionless distance y^+ from the wall for the universal velocity profile.

rise to eddy diffusivity variations which are continuous in both value and gradient. Here we will continue to work with the universal velocity profile.

19.3 Heat transfer in turbulent flow

For heat transfer across a laminar film, we write, as usual, that the heat flux ϕ is given by molecular conduction only

$$\phi = -k\frac{\mathrm{d}T}{\mathrm{d}y}, \tag{19.15}$$

where k is the molecular thermal conductivity (W/m K), and $\mathrm{d}T/\mathrm{d}y$ is the temperature gradient (K/m). As for momentum transport, turbulence in the fluid increases the heat flux more than would be expected on the basis of the change in temperature gradient. We can write

$$\phi = -(k + \rho\varepsilon C_\mathrm{p})\frac{\mathrm{d}T}{\mathrm{d}y}, \tag{19.16}$$

where the term in brackets is the effective thermal conductivity. The purely molecular value is enhanced by the turbulent transport. By using the eddy diffusivity again and, by implication, using the same value as was used for momentum transport, it is being assumed that a type of Reynolds analogy between momentum transport and thermal energy transport holds (see Hewitt 1961).

It is convenient to make the eqn (19.16) dimensionless by defining T^+, a dimensionless temperature, as

$$T^+ = (T_\mathrm{w} - T)C_\mathrm{pl}\frac{(\tau\rho)^{\frac{1}{2}}}{\phi}, \tag{19.17}$$

where T_w is the wall temperature (K). Thus $T^+ = 0$ at the wall, which corresponds to y or $y^+ = 0$. Substituting eqns (19.17) and (19.4) into eqn (19.16) gives a dimensionless version of the latter equation

$$\frac{dT^+}{dy^+} = \frac{1}{1/Pr + \rho\varepsilon/\mu} \tag{19.18}$$

or, using $\varepsilon^+ = \rho\varepsilon/\mu$

$$\frac{dT^+}{dy^+} = \frac{1}{1/Pr + \varepsilon^+}, \tag{19.19}$$

where the Prandtl number Pr is defined using the liquid properties, as we are concerned with the temperature profile in the liquid

$$Pr = \frac{\mu_1 C_{pl}}{k_1}. \tag{19.20}$$

Equation (19.19) can now be integrated over the three regions of the universal velocity profile.

(a) For $y^+ < 5$, $\varepsilon^+ = 0$, therefore

$$\frac{dT^+}{dy^+} = Pr \tag{19.21}$$

and

$$T^+ = Pr\, y^+, \tag{19.22}$$

as $T^+ = 0$ when $y^+ = 0$.

(b) For $5 < y^+ < 30$, $\varepsilon^+ = y^+/5 - 1$, therefore

$$\frac{dT^+}{dy^+} = \frac{5}{5(1/Pr - 1) + y^+} \tag{19.23}$$

and

$$T^+ = 5\left[Pr + \ln\left\{ 1 + Pr\left(\frac{y^+}{5} - 1\right)\right\}\right], \tag{19.24}$$

where the constant of integration has been obtained from the value of T^+ at $y^+ = 5$ in eqn (19.22).

(c) For $y^+ > 30$, $\varepsilon^+ = y^+/2.5 - 1$. The algebra is tedious but the result is

$$T^+ = 2.5 \ln\left\{ \frac{y^+ + 2.5(1/Pr - 1)}{30 + 2.5(1/Pr - 1)}\right\} + 5[Pr + \ln(1 + 5Pr)], \tag{19.25}$$

where the constant of integration has this time been obtained from the value of T^+ at $y^+ = 30$ from eqn (19.24).

19.4 Example

In Chapter 4 an example was considered in which:

$d = 0.032$ m,
$W_{\mathrm{lf}} = 0.037$ kg/s,
$\rho_1 = 1000$ kg/m^3,
$\mu_1 = 10^{-3}$ N s/m^2, and
$\tau = 16.8$ N/m^2,

and it was found that $W_{\mathrm{lf}}^+ = 368$ and $y^+ = 36.1$. Now, taking $Pr = 2$ and using the equation for $y^+ > 30$, T^+ can be calculated to be

$$T^+ = 2.5 \ln\left\{\frac{36.1 + 2.5(-\tfrac{1}{2})}{30 + 2.5(-\tfrac{1}{2})}\right\} + 5[2 + \ln(1 + 5 \times 2)]$$

$$= 22.5.$$

This can be interpreted in terms of a heat transfer coefficient because, using the definition of T^+ (eqn (19.17))

$$h = \frac{\phi}{\Delta T} = \frac{\phi}{T_{\mathrm{w}} - T} = \frac{C_{\mathrm{pl}}(\tau\rho)^{\frac{1}{2}}}{T^+} \tag{19.26}$$

and, in this case,

$$h = \frac{4200 \times (16.8 \times 1000)^{\frac{1}{2}}}{22.5} = 24\,000 \text{ W/m}^2 \text{ K},$$

where a value of 4200 J/kg K has been taken for C_{pl}. It can also be seen what influence turbulence has had on this heat transfer coefficient. If the film were laminar then, from Chapter 4, (eqn (4.17)), $W_{\mathrm{lf}}^+ = \tfrac{1}{2}y^{+2}$. Here, $W_{\mathrm{lf}}^+ = 368$, so $y^+ = 27.1$. For heat transfer in the 'laminar' layer, $\varepsilon^+ = 0$ so, from eqn (19.22),

$$T^+ = Pr\,y^+ = 2 \times 27.1 = 54.2$$

and then, using eqn (19.26), the heat transfer coefficient is

$$h = \frac{C_{\mathrm{p}}(\tau\rho)^{\frac{1}{2}}}{T^+} = \frac{4200 \times (16.8 \times 1000)^{\frac{1}{2}}}{54.2} = 10\,000 \text{ W/m}^2 \text{ K}.$$

Thus the turbulence has increased the heat transfer rate by a factor of about 2.4.

19.5 Does nucleate boiling or convective boiling occur?

A tentative answer can now be given to the question as to whether nucleate boiling or convective boiling takes place. Using the Rohsenow

(1952) correlation for nucleate boiling (or Mostinski 1963; or Cooper 1984), then a relationship between the temperature difference and heat flux in nucleate boiling can be established. Suppose, with reference to the above example, that this nucleate boiling equation is found to be, using SI units,

$$\Delta T_{sat} = 0.2\phi^{0.33}. \tag{19.27}$$

Convective boiling gave a heat transfer coefficient of 24 000 W/m² K, and therefore

$$\Delta T_{sat} = \phi/24\,000. \tag{19.28}$$

In eqns (19.27) and (19.28), the units which must be used are

$$\Delta T_{sat} \quad K$$
$$\phi \quad W/m^2.$$

Plotting the two equations for nucleate boiling and convective boiling on the same axes we obtain Fig. 19.2. The curves cross at values of

$\Delta T_{sat} = 13.9$ K, and
$\phi \quad = 330\,000$ W/m².

The boiling mechanism which actually occurs is that which:

(*a*) for a given value of ΔT_{sat} gives higher values of ϕ, or
(*b*) for a given value of ϕ gives lower value of ΔT_{sat} or, more generally,
(*c*) gives the higher heat transfer coefficient.

As remarked before with regard to the Chen correlation for flow boiling in Chapter 16, the mechanism does not suddenly switch, but rather is a gradual process.

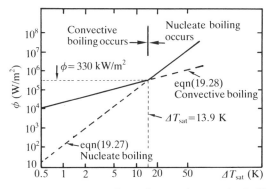

Fig. 19.2. Boiling curves for nucleate and convective boiling.

19.6   Comparison of predicted heat transfer coefficients with experimental values

Surprisingly, direct comparison with experiment is rather limited. One reason for this is that boiling heat transfer coefficients in the convective region are always high and are rarely the controlling resistance. Some low-pressure steam–water results have, however, been given by Aounallah *et al.* (1982). These experiments were performed in the apparatus shown in Fig. 19.3(a). The lower heated section generated the two-phase flow which was then allowed to come towards hydrodynamic equilibrium in the unheated section. The top heated section was where the measurements were taken. By looking particularly at the lower part of the top heated section, the heat flux could be varied almost independently of the quality. By this means it was found that the heat transfer coefficient was almost independent of the heat flux, but was determined by the mass flux and the quality. The results are shown in Fig. 19.3(b). The prediction of the heat transfer coefficient is almost always too high. This is particularly true at the higher mass flux and at higher qualities.

One suggestion for the reason for this overprediction relates to the difference between flow in a film (as in annular flow) and flow near the wall in a single-phase flow (see Fig. 19.4). Although many features of the velocity profile are similar in the region near the wall, the main difference between these two cases is the presence of the gas–liquid interface in the film-flow case. The surface tension at the interface may serve to damp out some of the turbulent fluctuations near the interface, and so it is possible that this 'interfacial damping' may reduce the eddy diffusivity in the immediate region of the interface. The overall effect of this would be to

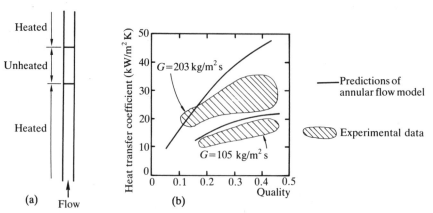

Fig. 19.3. Comparison of calculated convective heat transfer coefficients with experimental results.

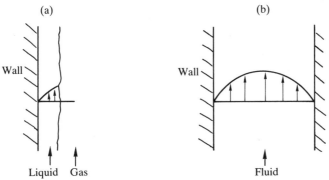

Fig. 19.4. Analogy between (a) turbulent film flow, and (b) turbulent single-phase flow.

reduce the predicted values of the heat transfer coefficients. However, this is still a matter of some doubt; further work will be necessary to clarify the situation.

19.7 Critical heat flux in steady flow with constant properties

The simplest case to consider is that for steady, rather than transient, flow and with constant physical properties so, for example, the variation of gas density with pressure is neglected. The attraction of the latter assumption is that the evaporation of the liquid is then entirely caused by the heat input, and there is no flashing of liquid to form vapour as the pressure falls. As earlier for adiabatic flow (see § 10.1) it is possible to perform mass balances on the gas, on the liquid film, and on the entrained liquid. The evaporation of the liquid film to produce gas now has to be taken into account (see Fig. 19.5).

$$\frac{\mathrm{d}G_g}{\mathrm{d}z} = \frac{4}{d}\frac{\phi}{\lambda}, \tag{19.29}$$

$$\frac{\mathrm{d}G_{lf}}{\mathrm{d}z} = \frac{4}{d}\left(D - E - \frac{\phi}{\lambda}\right), \tag{19.30}$$

and

$$\frac{\mathrm{d}G_{le}}{\mathrm{d}z} = \frac{4}{d}(E - D), \tag{19.31}$$

where ϕ is the heat flux at the wall (W/m^2), λ is the latent heat of vaporization (J/kg), E is the entrainment flux (per unit area of tube wall) (kg/m^2 s), D is the deposition flux (per unit area of tube wall) (kg/m^2 s), and d is the tube diameter (m). Note that

$$\frac{\mathrm{d}G_l}{\mathrm{d}z} = \frac{\mathrm{d}(G_{lf} + G_{le})}{\mathrm{d}z} = -\frac{4}{d}\frac{\phi}{\lambda}, \tag{19.32}$$

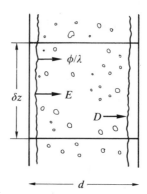

Fig. 19.5. Element of tube for mass balances.

and so

$$\frac{dG_l}{dz} + \frac{dG_g}{dz} = 0. \tag{19.33}$$

Hence, overall, mass is conserved (as must be the case).

If the heat flux is known, as a function of axial distance z if necessary, then eqn (19.29) can be integrated easily. In order to find the film flow rate at any axial point it is necessary to integrate eqn (19.30), which is much more difficult. From § 9.6 and § 9.7 we can put

$$D = kC \tag{19.34}$$

and

$$E = kC_E, \tag{19.35}$$

then

$$\frac{dG_{lf}}{dz} = \frac{4}{d}\left[k(C - C_E) - \frac{\phi}{\lambda}\right], \tag{19.36}$$

Again, as in adiabatic flow (see § 10.3),

$$C = \frac{G_{le}}{G_g/\rho_g + G_{le}/\rho_l} \tag{19.37}$$

and

$$C_E = f(\tau_i m/\sigma), \tag{19.38}$$

where m is the mean film thickness (m). It should be noted that these equations are just as for the adiabatic case, only the evaporation term ϕ/λ is additional.

Two problems now remain, apart from purely numerical problems in integrating the equation.

(a) *The value of the droplet mass transfer coefficient k.* This has been measured for 70 bar steam–water flow and has been found to be

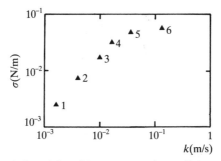

Fig. 19.6. Assumed variation of deposition mass transfer coefficient with surface tension.

0.01 m/s, whereas (see § 9.6) a value of 0.15 m/s for air–water was obtained. The reason for the low value in high-pressure steam–water is not clear. It may be a result of the high gas density or a result of the changed drop size which, to some extent, is controlled by the surface tension. The value has been estimated for other fluid systems and pressures by finding by trial and error the value necessary to give good CHF predictions. The resulting mass transfer coefficients have been correlated with surface tension (see Fig. 19.6 and Table 19.1) although, as noted above, this masks more complicated effects (see Whalley *et al.* 1974).

TABLE 19.1. *Pressures and fluids used in Fig. 19.6.*

Data point	Fluid	p (bar)
1	Steam–water	180
2	R12	10.5
3	Steam–water	69
4	Steam–water	24
5	Steam–water	10
6	Steam–water	1.5

(*b*) *Boundary conditions.* Where in the flow (at what quality) should the integration begin, and what there are the initial conditions? By trial and error it was found that an initial quality of 1 % and an initial fraction of liquid entrained of 99 % gave the best results.

19.8 Example

Take, as an example, a steam–water flow at 30 bar, $d = 15$ mm, $L = 5$ m, $G = 1000$ kg/m^2 s, $\Delta h_s = 0$. For this case $k = 0.015$ m/s and the CHF value

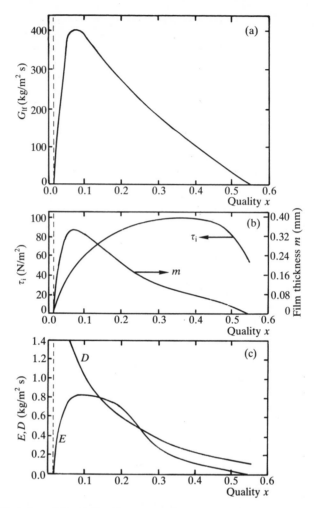

Fig. 19.7. Variation of liquid film flow rate and other parameters at the critical heat flux (from Whalley *et al.* 1974).

which is calculated is $\phi_c = 766 \, \text{kW/m}^2$. This is the smallest value of ϕ which just gives dryout, signified by a value of G_{lf} of zero at the top of the tube. These results are plotted against quality in Fig. 19.7. Figure 19.7(a) shows the variation of film flow rate, Fig. 19.7(b) shows the variation of mean film thickness and interfacial shear stress, and Fig. 19.7(c) shows the variation of entrainment and deposition rates. A number of points can be noted from this example and the results as shown in Fig. 19.7.

(*a*) The liquid film flow rate initially rises very steeply, this is because the deposition rate is very large and, because the quality is low, the flow velocity of the mixture is very low.

(*b*) The overall pressure gradient is more or less the same shape as the interfacial shear stress variation. In particular, it has been found experimentally that the shear stress and the pressure gradient drop just before CHF occurs. This can now be seen to be due to the thinning of the liquid film prior to the occurrence of dryout and, in particular, to the effect this has on the interfacial friction factor (see § 4.5).

(*c*) At no point in the flow can the entrainment or the deposition be neglected in comparison to the evaporation term ϕ/λ which, in this case, has a value of $0.43 \, \mathrm{kg/m^2 \, s}$.

By putting the heat flux to be a function of axial distance z, the cold-patch experiments referred to in § 17.3 can be successfully predicted (see Hewitt and Hall Taylor 1970).

19.9 Application to flow along rod bundles

Bundles of heated rods are commonly used in nuclear reactors. The rods contain the nuclear fuel, and the coolant flows between the rods. Some reactors are cooled by a boiling fluid, as in the Boiling Water Reactor; others are cooled by single-phase liquid, as in the Pressurized Water Reactor and the CANDU system, which only boils in the event of an accident. Taking a simple bundle with seven rods inside a shroud tube, as in Fig. 19.8(a), the flow area can be divided into three types of subchannel:

type 1 containing the centre rod,
type 2 containing one outer rod, and
type 3 adjacent to the shroud tube.

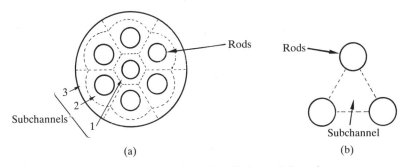

Fig. 19.8. Division of a rod bundle into subchannels.

These are *rod-centered* subchannels, that is each rod is in the centre of a subchannel. The alternative is to form subchannels by joining rod centres, as in Fig. 19.8(b). The advantage of a rod-centred subchannel is that a rod, with its liquid film, is entirely contained in one subchannel. Equations of the type used for a single tube are then used for each subchannel type. Now

$$\frac{\mathrm{d}G_{\mathrm{lf}}}{\mathrm{d}z} = f\left(D, E, \frac{\phi}{\lambda}\right) \tag{19.39}$$

as before for a single tube (see eqn (19.30)), but the diameter is replaced by a hydraulic diameter d_{h} which is given by

$$d_{\mathrm{h}} = \frac{4 \times \text{flow area}}{\text{wetted perimeter}}. \tag{19.40}$$

However, the equation for the entrained liquid mass flux has to be modified considerably

$$\frac{\mathrm{d}G_{\mathrm{le}}}{\mathrm{d}z} = f(D, E) + \text{term due to cross flow of vapour} \\ \text{between subchannels}$$

$$+ \text{term due to turbulent mixing} \\ \text{between subchannels.} \tag{19.41}$$

The vapour is assumed to flow between subchannels in order to make the axial pressure gradient in each subchannel the same. Thus vapour will move from a subchannel with a high heat flux to a subchannel with a low heat flux, carrying with it entrained liquid drops. Thus in Fig. 19.8(a) there will be a vapour flow from the heated subchannels 1 and 2 to the

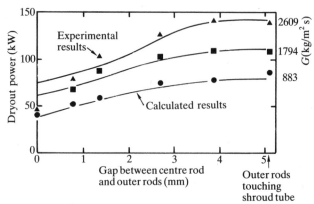

Fig. 19.9. Variation of dryout power with rod spacing for the rod bundle in Fig. 19.8(a) (from Whalley 1977).

unheated subchannel 3. The computational problems can be seen to be rather tedious. Some experimental and computed results are shown in Fig. 19.9 for a seven-rod bundle of the type illustrated in Fig. 19.8a. The fluid was Refrigerant-12 at a pressure of 10.7 bar. The special feature of these experiments was that the bundle geometry was varied: the outer rods initially touched the centre rod, then they were moved out in steps until they touched the shroud tube. The computed results are in good agreement with the calculated results, except when the gap between the centre rod and the outer tubes is very small. Perhaps this is not too surprising, as then subchannel type 2 (see Fig. 19.8) is very distorted and almost separated into two parts.

19.10 Transient flow

When a time variation is imposed on the system, for example variations of inlet flow, heat flux, or pressure with time, the mass fluxes become functions both of position and of time, so now we have partial differential equations instead of the ordinary differential equations (eqns (19.29–31)). For each phase (gas, liquid film, or liquid drops) a mass balance for an element of length δz and for the time interval δt (see Fig. 19.10) can be formed

$$\text{inflow} = \text{outflow} - \text{generation} + \text{accumulation}. \qquad (19.42)$$

For the gas, for example,

$$\text{inflow} = G_g \frac{\pi d^2}{4}\,\delta t, \qquad (19.43)$$

$$\text{outflow} = \left(G_g + \frac{\partial G_g}{\partial z}\,\delta z \right) \frac{\pi d^2}{4}\,\delta t, \qquad (19.44)$$

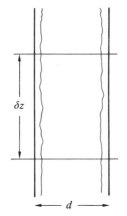

Fig. 19.10. Element of tube for mass balances.

and

$$\text{generation} = \pi d \delta z \left(\frac{\phi}{\lambda} + F \right) \delta t, \tag{19.45}$$

where the first term in the brackets represents vapour generation by direct evaporation and the second term represents vapour generation by flashing of liquid as the pressure changes. Also

$$\text{accumulation} = \frac{\pi d^2}{4} \delta z \frac{\partial}{\partial t} (\rho_g \alpha) \delta t. \tag{19.46}$$

Substituting into eqn (19.42) gives

$$G_g \frac{\pi d^2}{4} \delta t = \left(G_g + \frac{\partial G_g}{\partial z} \delta z \right) \frac{\pi d^2}{4} \delta t - \pi d \delta z \left(\frac{\phi}{\lambda} + F \right) \delta t,$$
$$+ \frac{\pi d^2}{4} \delta z \frac{\partial}{\partial t} (\rho_g \alpha) \delta t, \tag{19.47}$$

and then rearranging gives

$$\frac{\partial}{\partial t} (\alpha \rho_g) + \frac{\partial G_g}{\partial z} = \frac{4}{d} \left(\frac{\phi}{\lambda} + F \right). \tag{19.48}$$

Similar, but more complicated, equations can be obtained which involve G_{le} and G_{lf}. Note that now the total mass flux at values of z and t $(G_g + G_{le} + G_{lf})$ is no longer constant.

It is also possible to write an overall energy equation for the element, which will involve

$\partial(\text{enthalpy terms})/\partial z$ which represent the flow of energy in and out of the element,

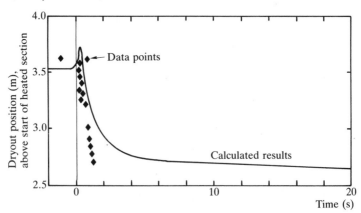

Fig. 19.11. Variation of dryout position with time during a fast depressurization (from Whalley *et al.* 1984).

∂(internal energy terms)$/\partial t$ which represent the storage of energy within the element, and

ϕ which represents the energy addition due to the heat flux

Typical results available are, for example, the variation of dryout position with time during a pressure transient (see Fig. 19.11). During this transient, the pressure fell from 65 bar to 12 bar in less than 2 s. The initial liquid mass flux was 1000 kg/m^2 s and increased temporarily by about 20 % during the transient.

19.11 Thermodynamic non-equilibrium effects

Another refinement which is possible, is to consider the combined effects of heat transfer and the variation of film and entrained flows. The heat transfer processes which must be taken into account are those from the wall to the liquid film, from the film to the vapour, and from the vapour to the liquid drops. Thus, the temperature of the various parts of the flow–liquid film, liquid drops, and vapour–can be found. It is necessary to know, or to be able to calculate, the size of drops and the droplet drag coefficient. This information is necessary to enable the droplet heat transfer coefficient to be calculated. Examples of the type of results it is then possible to produce are:

(a) the temperature of the wall, film, and vapour at various positions along the channel;

(b) the drop size and the number of drops, which go to make up the entrained liquid flow rate; and

(c) the liquid film flow rate.

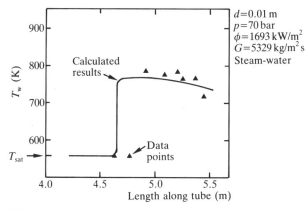

Fig. 19.12. Wall temperature rise at dryout: calculated and experimental results (from Whalley *et al.* 1982).

Some typical results are shown in Fig. 19.12. This is a case where the critical heat flux is exceeded part way along the tube. The large temperature rise, here about 200 °C, occurs at dryout and illustrates that heat transfer after dryout (post-dryout heat transfer) is relatively poor. The temperature rise found here would have been even larger if the experiments had been carried out at a lower pressure.

References

AOUNALLAH, Y., KENNING, D. B. R., WHALLEY, P. B., and HEWITT, G. F. (1982). Boiling heat transfer in annular flow. *Seventh Int. Heat Transfer Conference, Munich* **4**, 193–9.

COOPER, M. G. (1984). Saturation nucleate pool boiling—a simple correlation. *First UK National Heat Transfer Conference* (I. Chem. E. Symp. Series, No. 86) **2**, 785–93.

HEWITT, G. F. (1961). Analysis of annular two-phase flow: application of the Dukler analysis to vertical upward flow in a tube. *AERE-R3680*.

HEWITT, G. F. (1978). Critical heat flux in flow boiling. *Sixth Int. Heat Transfer Conference, Toronto* **6**, 143–71.

HEWITT, G. F. and HALL TAYLOR, N. S. (1970). *Annular two-phase flow*. Pergamon Press, Oxford.

MOSTINSKI, I. L. (1963). Calculation of heat transfer and critical heat flux in boiling liquids based on the law of corresponding states. *Teploenergetika* **10** (No. 4), 66–71.

ROHSENOW, W. M. (1952). A method of correlating heat transfer data for surface boiling of liquids. *Trans. ASME* **74**, 969.

WHALLEY, P. B., HUTCHINSON, P., and HEWITT, G. F. (1974). The calculation of critical heat flux in forced convection boiling. *Fifth Int. Heat Transfer Conference, Tokyo,* paper B6.11.

WHALLEY, P. B. (1977). The calculation of dryout in rod bundles. *Int. J. Multiphase Flow* **3**, 501–15.

WHALLEY, P. B., AZZOPARDI, B. J., HEWITT, G. F., and OWEN, R. G. (1982). A physical model of two-phase flow with thermodynamic and hydrodynamic non-equilibrium. *Seventh Int. Heat Transfer Conference, Munich* **5**, 181–8.

WHALLEY, P. B., LYONS, A. J., and SWINNERTON, D. (1984). Transient critical heat flux in flow boiling. *First UK National Heat Transfer Conference* (I. Chem. E. Symp. Series No. 86) **2**, 805–16.

20

POST-BURNOUT HEAT TRANSFER

20.1 Introduction

Heat transfer after the critical heat flux has been exceeded (*post-burnout heat transfer*) is relatively poor compared with pre-burnout heat transfer. At the critical heat flux point, the heat transfer coefficient may fall by a factor of up to 100. The general post-dryout heat transfer regime can be divided into two general types:

(*a*) film boiling which occurs, for example, in pool boiling or after the subcooled type of flow-boiling critical heat flux has occurred, (see § 17.2), and

(*b*) post-dryout flow boiling, which occurs after the high quality dryout type of flow-boiling critical heat flux has occurred (see § 17.3).

20.2 Film boiling

The important feature of film boiling is that there is a thin film of vapour adjacent to the heated surface (see Fig. 20.1(a)). The heat is conducted through the vapour film, and evaporation takes place at the liquid–vapour interface. This will be seen later to be completely analogous to filmwise condensation and, by analogy, the average heat transfer coefficient h, if the plate is of height L, is

$$h = \frac{2\sqrt{2}}{3}\left[\frac{\rho_g(\rho_l - \rho_g)g\lambda k_g^3}{\mu_g \Delta T_{sat} L}\right]^{\frac{1}{4}}. \tag{20.1}$$

For film boiling outside a cylinder of diameter d (see Figure 20.1(b))

$$h = 0.62\left[\frac{\rho_g(\rho_l - \rho_g)g\lambda k_g^3}{\mu_g \Delta T_{sat} d}\right]^{\frac{1}{4}}, \tag{20.2}$$

where the constant (0.62) is slightly different from the one in the condensation case (which is 0.73) on the basis of experimental evidence (see Bromley 1950). The physical properties are ideally evaluated at the mean temperature of the film, that is at a temperature

$$T = T_w - \frac{\Delta T_{sat}}{2}. \tag{20.3}$$

191

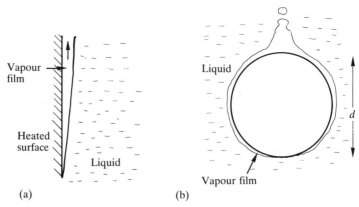

Fig. 20.1. Film boiling (a) on a vertical flat plate and (b) on a horizontal cylinder.

Taking an example for steam–water at 1 bar, for which

$$T_{\text{sat}} = 100\,°\text{C},$$
$$T_{\text{w}} = 700\,°\text{C},$$
$$\Delta T_{\text{sat}} = 600\,°\text{C},$$
$$\lambda = 2256 \times 10^3\,\text{J/kg},$$
$$k_{\text{g}} = 0.055\,\text{W/m K},$$
$$\mu_{\text{g}} = 24 \times 10^{-6}\,\text{N s/m}^2,$$
$$\rho_{\text{g}} = 0.32\,\text{kg/m}^3,$$
$$\rho_{\text{l}} = 1000\,\text{kg/m}^3, \quad \text{and}$$
$$d = 0.02\,\text{m},$$

where the physical properties above have been evaluated at 400 °C. Then from eqn (20.2)

$$h = 157\,\text{W/m}^2\,\text{K}$$

and so

$$\phi = h\Delta T_{\text{sat}} = 94\,000\,\text{W/m}^2,$$

whereas the critical heat flux for water at 1 bar is around 1.25 MW/m² (see § 15.7). So, although the tube is very hot, it is carrying only a fraction of the critical heat flux. So we are around point A on the boiling curve (see Fig. 20.2). The point C, which has the same heat flux as at point B, can be found. The temperature at C will be so high that radiative heat transfer is very important, hence

$$h_{\text{total}} = h_{\text{film boiling}} + a h_{\text{radiative}}, \tag{20.4}$$

where

$$h_{\text{radiative}} = \frac{\varepsilon\sigma(T_{\text{w}}^4 - T_{\text{sat}}^4)}{T_{\text{w}} - T_{\text{sat}}}, \tag{20.5}$$

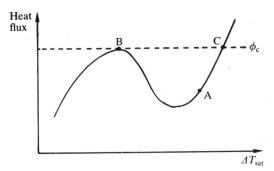

Fig. 20.2. Boiling curve.

and where ε is the surface emissivity, and σ is Stefan's constant $(56.7 \times 10^{-9}\,\mathrm{W/m^2\,K^4})$.

The empirical constant a has been found by experiment to have a value of $\frac{3}{4}$. The use of the constant a in eqn (20.4) is an attempt to allow for the effect of the radiative heat transfer on the film.

In the presence of an upward liquid velocity u across the tube (see Fig. 20.3(a)), the film boiling heat transfer coefficient is given by (Bromley *et al.* 1953)

$$h = 2.7 \left(\frac{\rho_g u \lambda k_g}{d \Delta T_{\text{sat}}} \right)^{\frac{1}{2}}, \qquad (20.6)$$

and now

$$h_{\text{total}} = h_{\text{film boiling}} + \tfrac{7}{8} h_{\text{radiative}}, \qquad (20.7)$$

where the factors 2.7 and 7/8 in these equations are experimentally determined. The condition for the velocity to be important can be obtained by saying that the dynamic pressure in the liquid $(\rho_1 u^2 / 2)$ should be greater than the hydrostatic pressure head across the tube

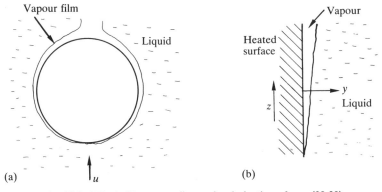

Fig. 20.3. Film boiling: co-ordinates for derivation of eqn (20.20).

$(\rho_l g d)$, hence

$$\frac{\rho_l u^2}{2} > \rho_l g d, \qquad (20.8)$$

or

$$\frac{u^2}{gd} > 2. \qquad (20.9)$$

This is sometimes quoted as:

if $u^2/gd > 4$ then the liquid velocity is definitely important and eqn (20.6) should be used; or

if $u^2/gd < 1$ then the liquid velocity is definitely not important and eqn (20.2) should be used.

The general form of eqn (20.6) can be obtained by a derivation similar to that for the Nusselt film condensation equation (see § 22.2.)

As with the Nusselt equation, the derivation is easiest for the case of a flat vertical plate (see Fig. 20.3(b)). If it is assumed that there is a constant shear stress in the vapour film, and that the vapour flow is laminar, then the velocity profile in the vapour film is

$$u = \frac{\tau}{\mu_g} y, \qquad (20.10)$$

where τ is the shear stress in the film (N/m^2), and u is the velocity in the film (m/s) at a distance y from the wall. The mass flow rate per unit width M of the film (kg/ms) is

$$M = \rho_g \int_0^\delta u \, dy = \frac{\rho_g \tau \delta^2}{2\mu_g}. \qquad (20.11)$$

At the outside edge of the film ($y = \delta$) then, from eqn (20.10),

$$\frac{\tau}{\mu_g} = \frac{u}{\delta} \qquad (20.12)$$

and so eqn (20.11) becomes

$$M = \frac{\rho_g u \delta}{2}. \qquad (20.13)$$

Now, as the film thickness changes, so does the mass flow rate in the film

$$\frac{dM}{dz} = \frac{\rho_g u}{2} \frac{d\delta}{dz}. \qquad (20.14)$$

However, the reason why mass flow rate and the film thickness change is because of the heat transfer, so

$$\phi = \lambda \frac{dM}{dz} = \frac{k_g}{\delta} \Delta T_{sat}. \tag{20.15}$$

The final part of eqn (20.15) is derived assuming that the heat is simply conducted across the vapour film. So, from eqns (20.14) and (20.15),

$$\delta d\delta = \frac{2k_g \Delta T_{sat}}{\rho_g u \lambda} dz \tag{20.16}$$

and integrating, since $\delta = 0$ when $z = 0$,

$$\delta = 2 \left(\frac{k_g \Delta T_{sat} z}{\rho_g u \lambda} \right)^{\frac{1}{2}}. \tag{20.17}$$

The local heat transfer coefficient h_z at z is then

$$h_z = \frac{k_g}{\delta} = \frac{1}{2} \left(\frac{\rho_g u \lambda k_g}{\Delta T_{sat} z} \right)^{\frac{1}{2}}. \tag{20.18}$$

The mean heat transfer coefficient h over a length L is given by

$$h = \frac{1}{L} \int_0^L h_z \, dz, \tag{20.19}$$

and thus

$$h = \left(\frac{\rho_g u \lambda k_g}{\Delta T_{sat} L} \right)^{\frac{1}{2}}, \tag{20.20}$$

which is of the form of eqn (20.6). Example 9 in Chapter 25 gives detailed information about the calculation of heat transfer coefficients in film boiling for cases with and without a liquid flow and with and without the radiation correction.

20.3 Post-burnout flow boiling

This is sometimes called the *liquid deficient region*; a typical idealized picture of the flow is shown in Fig. 20.4. Typically, the wall temperature reaches a peak and then falls as shown in Fig. 20.5. In this figure dryout occurs at point A and the last of the liquid drops evaporate at point B.

A number of types of procedure and correlation have been used to calculate the heat transfer rate (typically by calculating the wall temperature). The main types are:

(a) purely empirical correlations;
(b) correlations which attempt to calculate the degree of thermodynamic non-equilibrium; and

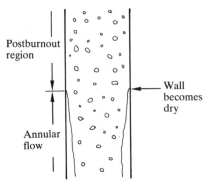

Fig. 20.4. High quality post-burnout flow.

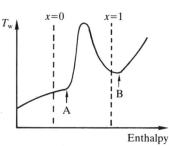

Fig. 20.5. Boiling curve for high quality post-burnout flow.

(c) theoretical models which attempt to model the different stages of the heat transfer process.

Examples of these calculation methods are given in the following sections.

20.4 Empirical correlations

For single-phase convective heat transfer it is usual to write

$$Nu = a\,Re^b\,Pr^c,\tag{20.21}$$

and for turbulent flow in smooth tubes the Dittus–Boelter equation corresponds to:

$$a = 0.023,$$
$$b = 0.8,\quad\text{and}$$
$$c = 0.4.$$

The two-phase flow is mostly gas, so the gas properties are used (at the wall temperature), so that

$$Nu = Nu_g = \frac{hd}{k_g}\tag{20.22}$$

and

$$Pr = Pr_g = \frac{\mu_g C_{pg}}{k_g}.\tag{20.23}$$

The Reynolds number used is defined by

$$Re = Re_g = \frac{u_h d \rho_g}{\mu_g},\tag{20.24}$$

where u_h is the homogeneous velocity

$$u_h = \frac{G}{\rho_h} = G\left(\frac{x}{\rho_g} + \frac{1-x}{\rho_l}\right)\tag{20.25}$$

and so

$$Re = \frac{Gd}{\mu_g}\left[x + \frac{\rho_g}{\rho_l}(1-x)\right] = Re_g\left[x + \frac{\rho_g}{\rho_l}(1-x)\right], \qquad (20.26)$$

where Re_g is the single-phase gas Reynolds number.

It can be noted that the definition of the Reynolds number (eqn (20.24)) is not completely logical because the homogeneous velocity is used in conjunction with the gas phase density. An additional multiplying factor Y was necessary

$$Y = 1 - 0.1\left[\left(\frac{\rho_l}{\rho_g} - 1\right)(1-x)\right]^{0.4}, \qquad (20.27)$$

and so finally

$$Nu_g = a\left\{Re_g\left|x + \frac{\rho_g}{\rho_l}(1-x)\right|\right\}^b Pr_g^c Y^d. \qquad (20.28)$$

It was found for a best fit with the experimental data that:

$$a = 0.0033,$$
$$b = 0.9,$$
$$c = 1.32, \quad \text{and}$$
$$d = -1.5.$$

The Prandtl number exponent c is quite different from the single-phase value. However, this is perhaps not very significant because, as can be seen below, the available Prandtl number range for the post-burnout data is quite small.

The above values of a, b, c, and d are for water in tubes and annuli for the following range of conditions (Groeneveld 1973):

2.5 mm $< d$ (or equivalent diameter) < 25 mm,
34 bar $< p < 215$ bar,
700 kg/m^2 s $< G < 5300$ kg/m^2 s,
$0.1 < x < 0.9$,
120 kW/m$^2 < \phi < 2100$ kW/m^2,
$95 < Nu_g < 1770$,
$6.6 \times 10^4 < Re_g[x + (1-x)\rho_g/\rho_l] < 1.3 \times 10^6$,
$0.88 < Pr_g < 2.21$, and
$0.71 < Y < 0.98$.

Empirical correlations such as eqn (20.28) give very good results but they cannot, of course, be used outside the range of the original data.

20.5 Departure from non-equilibrium

The fact that superheated gas and liquid drops coexist means that there is some thermodynamic non-equilibrium. Clearly, two extremes are, in theory, possible.

(a) *Complete equilibrium*. The gas temperature therefore remains at the saturation temperature until the last liquid drop has evaporated. This is most nearly true at high pressures and high flow rates. In this case the pressure is high when it approaches the critical pressure and the mass flux is high when it exceeds $3000 \, \text{kg/m}^2 \, \text{s}$.

(b) *Complete non-equilibrium*. All the heat goes into raising the gas temperature and no droplets evaporate. This is most nearly true at low pressure and low flow rates. In this case the heat transfer coefficient for the gas flow can be calculated from single-phase results.

One way to achieve an intermediate position between these two extremes is to write

$$\phi = \phi_g + \phi_l, \qquad (20.29)$$

where ϕ is the total heat flux (W/m^2), ϕ_g is the heat flux to the gas (W/m^2) (that is, the flux which raises the gas temperature), ϕ_l is the heat flux to liquid (W/m^2) (that is, the flux which leads to the evaporation of liquid droplets); and then to put

$$\beta = \frac{\phi_l}{\phi_g}. \qquad (20.30)$$

Then β can be assumed to be a function of position, or quality, or to be a constant. Assuming β to be a constant is obviously very simple, but it leads to linear variations of gas temperature and quality, not to the curves which are actually found (see Collier 1981).

20.6 Theoretical models

Various levels of complexity for theoretical models are possible. A simple theoretical model (see, for example, Whalley *et al.* 1982) would include the effects of:

(a) heat transfer from the wall to the gas (this can be done simply, by assuming that the droplet laden gas behaves like a single-phase gas, so that in this case the droplets are ignored); and

(b) heat transfer from the superheated gas to the droplets, resulting in some evaporation.

In this simple model the effects of heat transfer from the wall directly to droplets which hit the wall is ignored, as are the effects of radiative heat transfer. Even this 'simple' model can give rise to considerable complexities. For example, to calculate the heat transfer to the drops from the gas

$$Nu = 2 + 0.6Re^{\frac{1}{2}}Pr_{g}^{\frac{1}{3}}, \tag{20.31}$$

where:

$$Nu = \frac{hD_d}{k_g}; \tag{20.32}$$

$$Re = \frac{\rho_g D_d(u_g - u_d)}{\mu_g}; \tag{20.33}$$

$$Pr_g = \frac{\mu_g C_{pg}}{k_g}; \tag{20.34}$$

D_d is the droplet diameter (m);
u_g is the gas velocity (m/s); and
u_d is the droplet velocity (m/s).

The first term of the right-hand side of eqn (20.31) is the conduction term and the second term on the right-hand side is the convection term. Hence the heat transfer rate to each drop is

$$Q = \pi D_d^2 h(T_g - T_d), \tag{20.35}$$

where T_g is the gas temperature (K), and T_d is the droplet temperature (K) which is assumed to be the saturation temperature.

In order to calculate Re, the droplet velocity u_d must be known. This must be calculated from a force balance on the drop. Assuming that the drop is not accelerating then the drag force is equal to the gravity force on the drop. The drag force F is calculated from

$$C_D = \frac{F/\frac{1}{4}\pi D_d^2}{\frac{1}{2}\rho_g(u_g - u_d)^2}, \tag{20.36}$$

where the drag coefficient C_D is given by

$$C_D = \frac{24}{Re} + 0.44. \tag{20.37}$$

Here the first term on the right-hand side is the laminar flow value and the second is the turbulent value. Adding the terms in this way gives a reasonable value of the drag coefficient at all Reynolds numbers. Then, as the drag force and the gravity force are equal,

$$F = \frac{\pi D_d^3}{6}(\rho_1 - \rho_g)g. \tag{20.38}$$

The whole calculation is a marching process along the tube. One point in the calculation procedure is illustrated in detail in Example 10 in Chapter 25. In order to start the process, some initial droplet size and population must be known. One suggestion for the initial drop size D_d has been that

$$\frac{D_d}{d} = 1.9 Re_g^{0.1} We_g^{-0.6} \left(\frac{\rho_g}{\rho_l}\right)^{0.6}, \qquad (20.39)$$

where

$$Re_g = \frac{Gxd}{\mu_g} \qquad (20.40)$$

and

$$We_g = \frac{G^2 x^2 d}{\rho_g \sigma}. \qquad (20.41)$$

Because the gas velocity is increasing there is the possibility that the drops are no longer stable. This tends to occur when the droplet Weber number We_d exceeds 7.5, in this case

$$We_d = \frac{\rho_g (u_g - u_d)^2 D_d}{\sigma}, \qquad (20.42)$$

and then the droplet is assumed to split into two equal-sized drops. In this way it can be seen that even the very simplest theoretical model quickly becomes very complicated, although very detailed results can be generated about the drop size distribution, wall temperature, drop velocities, drop temperature, gas temperature, and droplet flux. However, all of these (except wall temperature) are difficult to measure. The gas temperature, for example, is difficult to measure because a thermocouple in the flow becomes wet due to the impact of drops and gives a result approximating to the drop temperature, which is not far from the saturation temperature. Thus, testing this type of model is difficult because only the wall temperature is known with any accuracy.

References

Bromley, L. A. (1950). Heat transfer in stable film boiling. *Chem. Eng. Prog.* **46**, 221–7.

Bromley, L. A., LeRoy, N. R., and Robbers, J. A. (1953). Heat transfer in forced convection film boiling. *Ind. Eng. Chem.* **49**, 1921–8.

Collier, J. G. (1981). *Convective boiling and condensation* (2nd edn). McGraw-Hill, New York.

Groeneveld, D. C. (1973). Post dryout heat transfer at reactor operating conditions. *AECL-4513* (ANS topical meeting on Water Reactor Safety, Salt Lake City).

Whalley, P. B., Azzopardi, B. J., Hewitt, G. F., and Owen, R. G. (1982). A physical model of two-phase flow with thermodynamic and hydro-dynamic non-equilibrium. *Seventh Int. Heat Transfer Conference, Munich* **5**, 181–8.

21

REWETTING OF HOT SURFACES

21.1 Introduction

It is well known that a liquid does not wet a hot surface. For example, a drop of water on a hot, horizontal plate will 'run' around in a chaotic manner and evaporate only slowly. This occurs because the liquid is separated from the plate by a thin film of vapour (see Fig. 21.1) so that the friction for sideways motion of the drop is very small and the heat transfer across the vapour film is poor. The vapour film, of course moves outwards, and fresh vapour is generated by evaporation at the underside of the drop due to heat conduction across the film and radiation from the plate to the drop. If the plate is allowed to cool down, it will eventually reach a temperature at which the vapour film collapses, and then very intense boiling takes place which rapidly leads to the evaporation of all the liquid. The surface temperature at which this sudden wetting of the plate occurs is the Leidenfrost temperature (Leidenfrost first observed the phenomenon in 1756; see Bergles 1981).

21.2 Leidenfrost temperature

The Leidenfrost phenomenon can be studied in the same way that pool-boiling critical heat flux was analysed: by considering the stability of the vapour film. The result is usually quoted in terms of the minimum heat flux ϕ_{min} which will maintain film boiling, as shown on the boiling curve in Fig. 21.2.

The result is of the form (see Zuber and Tribus 1958)

$$\phi_{min} = C\lambda\rho_g\left[\frac{\sigma g(\rho_1 - \rho_g)}{(\rho_1 + \rho_g)^2}\right]^{\frac{1}{4}}, \tag{21.1}$$

where C is a non-dimensional constant which lies between 0.09 and 0.18; 0.13 is sometimes taken as an intermediate value.

So, for example, for steam–water at 1 bar, the relevant physical properties are: $\lambda = 2256 \times 10^3$ J/kg, $\sigma = 0.059$ N/m, $\rho_g = 0.598$ kg/m³, and $\rho_1 = 1000$ kg/m³. Hence, using a value for C of 0.13, the calculated value of ϕ_{min} is 27 kW/m². This can be compared with the pool-boiling critical heat flux for steam–water, which at 1 bar is 1.25 MW/m². The physical properties above have been taken to be the values at the

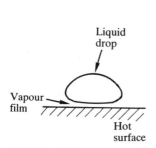

Fig. 21.1. Liquid drop on a hot surface.

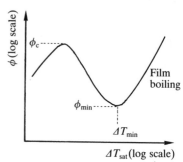

Fig. 21.2. Boiling curve showing minimum film boiling point.

saturation temperature (100 °C), rather than at some mean film temperature.

The value of ΔT_{sat} at ϕ_{min} is ΔT_{min} (see Fig. 21.2). This can now be calculated using the film-boiling equation from § 20.2

$$h = 0.62 \left[\frac{\rho_g(\rho_1 - \rho_g)g\lambda k_g^3}{\mu_g \Delta T_{sat} d} \right]^{\frac{1}{4}}. \tag{21.2}$$

For a horizontal plate d must be replaced by another length scale l:

$$l = 2\pi \left[\frac{\sigma}{g(\rho_1 - \rho_g)} \right]^{\frac{1}{2}} \tag{21.3}$$

is often used.

In the above example $\rho_g = 0.598 \, \text{kg/m}^3$, and so $l = 0.015 \, \text{m}$. Then, taking $k_g = 0.025 \, \text{W/m K}$ and $\mu_g = 12 \times 10^{-6} \, \text{N s/m}^2$ (values at 100 °C),

$$h = \frac{\phi_{min}}{\Delta T_{min}} = \frac{1010}{\Delta T_{min}^{\frac{1}{4}}}. \tag{21.4}$$

Since $\phi_{min} = 27 \, \text{kW/m}^2$, then $\Delta T_{min} = 80 \, °C$. This is a rather low value, giving a minimum film boiling temperature of 180 °C. In practice, effects such as liquid and surface contamination and the presence of an oxide layer on the surface all tend to increase the minimum film boiling temperature, another term for the Leidenfrost temperature (see Bergles and Thompson 1970), and so the value of ϕ_{min} is also increased. Indeed, it seems probable that there is, in general, no well-defined minimum film-boiling heat flux and, correspondingly, no well-defined minimum film-boiling temperature. Nevertheless these quantities are calculated in Example 7 in Chaper 25 as part of the pool-boiling curve calculation. Typical values for the minimum film-boiling temperature for water at 1 bar are around 290 °C.

21.3 Completion of the boiling curve

The idealized boiling curve (see Fig. 21.3) may now be drawn. The regions and special points on the curve are now:

up to A	natural convection heat transfer;
A	onset of nucleate boiling (see § 14.7);
A to B	nucleate boiling heat transfer (see § 14.8);
B	critical heat flux (see § 15.4);
B to C	transition boiling;
C	minimum film-boiling temperature (see § 21.2); and
C onwards	film boiling (see § 20.2).

The transition boiling region is the main region not yet considered. Actually, comparatively little is known, and one assumption made is that the transition boiling line (B to C) is the straight line connecting B and C when plotted (as above) on a log–log graph. This is illustrated in Fig. 21.3 and in Example 7 in Chapter 25.

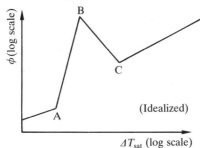

Fig. 21.3. Idealized pool boiling curve.

21.4 Vapour film thickness underneath the liquid drop

The vapour film thickness underneath a liquid drop on a horizontal plate (see Fig. 21.4(a)) can be calculated if a number of assumptions are made:

(a) vapour film thickness δ is constant;
(b) vapour flow outwards is laminar;
(c) heat transfer is by conduction only; and
(d) the drop is hemispherical.

Fig. 21.4. Liquid drop on a hot surface: co-ordinate system.

First consider laminar, one-dimensional flow between parallel plates (see Fig. 21.4(b)). The equation of motion is

$$\mu_g \frac{d^2 u}{dy^2} = \frac{dp}{dr}.$$ (21.5)

Now applying this to the radial flow by arguing that the radial velocity gradients are relatively small in comparison to the velocity gradients in the y direction gives

$$u_r = \frac{1}{\mu_g} \frac{dp}{dr} \frac{y^2}{2} + Ay + B,$$ (21.6)

where u_r is the radial velocity, which is a function of y. The boundary conditions are that:

$u_r = 0$ when $y = 0$, so $B = 0$; and
$u_r = 0$ when $y = \delta$, so $A = -\dfrac{1}{\mu_g} \dfrac{dp}{dr} \dfrac{\delta}{2}$

so that eqn (21.5) becomes

$$u_r = \frac{1}{\mu_g} \frac{dp}{dr} \frac{y}{2} (y - \delta),$$ (21.6)

and the mean velocity in the r direction $\bar{u}_r$ is given by

$$\bar{u}_r = \frac{1}{\delta} \int_0^\delta u \, dy = -\frac{\delta^2}{12} \frac{1}{\mu_g} \frac{dp}{dr}.$$ (21.7)

Now, evaluating the evaporation gas mass flux $\dot{m}_g$ (per unit area of the underside of the drop)

$$\phi = k_g \frac{\Delta T_{\text{sat}}}{\delta},$$ (21.8)

$$\dot{m}_g = \frac{\phi}{\lambda} = \frac{k_g \Delta T_{\text{sat}}}{\delta \lambda}.$$ (21.9)

Next, work out $\bar{u}_r$ at a radius r by equating the vapour generation rate from $r = 0$ to r to the vapour mass flow rate crossing $r = r$.

$$\pi r^2 \dot{m}_g = 2\pi r \bar{u}_r \delta \rho_g,$$ (21.10)

so that

$$\bar{u}_r = \frac{r \dot{m}_g}{2\rho_g \delta} = \frac{r k_g \Delta T_{\text{sat}}}{2\rho_g \lambda \delta^2},$$ (21.11)

and therefore, using eqn (21.7)

$$\frac{dp}{dr} = -\frac{12 \mu_g \bar{u}_r}{\delta^2} = -\frac{6 \mu_g k_g \Delta T_{\text{sat}} r}{\rho_g \lambda \delta^4}$$ (21.12)

Integrating eqn (21.12)

$$p = -\frac{6\mu_g k_g \Delta T_{sat}}{\rho_g \lambda \delta^4}\frac{r^2}{2} + C, \tag{21.13}$$

where the boundary condition is that $p = p_0$ (atmospheric pressure) at $r = R$ (the outer radius of the drop). Therefore

$$p - p_0 = \frac{3\mu_g k_g \Delta T_{sat}}{\rho_g \lambda \delta^4}(R^2 - r^2). \tag{21.14}$$

Finally, this pressure difference must be capable of supporting the drop

$$\int_0^R (p - p_0)2\pi r\, dr = \rho_1 g \tfrac{2}{3}\pi R^3. \tag{21.15}$$

Substituting from eqn (21.14), we obtain

$$\frac{6\pi \mu_g k_g \Delta T_{sat}}{\rho_g \lambda \delta^4}\int_0^R (R^2 r - r^3)\, dr = \rho_1 g \tfrac{2}{3}\pi R^3 \tag{21.16}$$

and, performing the integration,

$$\frac{6\pi \mu_g k_g \Delta T_{sat}}{\rho_g \lambda \delta^4}\frac{R^4}{4} = \rho_1 g \tfrac{2}{3}\pi R^3. \tag{21.17}$$

Rearranging, the final result for the film thickness is

$$\delta = \left(\frac{9}{4}\frac{\mu_g k_g \Delta T_{sat} R}{\rho_1 \rho_g g \lambda}\right)^{\frac{1}{4}}. \tag{21.18}$$

Now, ignoring the effect of temperature on the physical properties, and taking the values for steam–water at 1 bar and 100 °C, where:

$\mu_g = 12 \times 10^{-6}\,N\,s/m^2$,
$k_g = 0.025\,W/m\,K$,
$\rho_1 = 1000\,kg/m^3$,
$\rho_g = 0.598\,kg/m^3$, and
$\lambda = 2256 \times 10^3\,J/kg$;

then, for a drop of radius 1 mm $(R = 10^{-3}\,m)$ and $\Delta T_{sat} = 80\,°C$, eqn (21.18) gives the result that

$$\delta = 45\,\mu m.$$

This is a very thin vapour film, and this film eventually breaks down because of an instability which causes the liquid to touch the hot surface; then wetting occurs. For further information see Bolle and Moureau (1982).

21.5 Quenching and reflooding

Quenching is the term used to describe the cooling of very hot metal by a cold liquid, for example the water spray used to cool steel in a rolling mill or the water used to cool an overheated nuclear reactor core. A nuclear reactor may be cooled by liquid (water) sprays, or by pumping large quantities of water into the core from the top (*top flooding*) or from the bottom (*bottom flooding*), or both. Quenching by top flooding is illustrated in Fig. 21.5. Here the upper part of the hot rod has been quenched and is covered by a liquid film. However, the lower part is still very hot, and so the liquid film running down the wall 'splutters' off at the boundary between the quenched and the hot sections of the rod. This boundary is the quench front, which moves slowly down the rod. It is often found that the quench front moves downwards with a constant velocity u. Typical results are shown in Fig. 21.6: plotting $1/u$ against the initial metal temperature often gives straight lines. In these experiments (from Bennett *et al.* 1966), the lines ceased to be straight when $1/u < 4\,\text{s/m}$ (or when $u > 0.25\,\text{m/s}$).

Bottom flooding can take two forms depending on whether the water flow rate is large or small. When the water flow rate is small (see Fig. 21.7(a)), the situation is not unlike co-current upwards annular flow, except that some spluttering occurs at the quench front. However, at high water flow rates (see Fig. 21.7(b)) a flow pattern often named inverted annular flow is formed. In this case there is a liquid core surrounded by an unstable vapour film. The quench front then occurs relatively low down in the flow.

The phenomena of top and bottom flooding are obviously extremely important and extremely complicated. Two things are important if the progress of the quench front is to be predicted in any way.

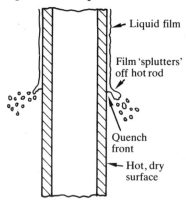

Fig. 21.5. Top flooding of a hot surface.

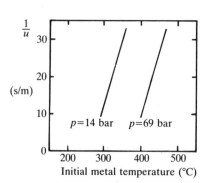

Fig. 21.6. Quenching front velocity in top flooding (from Bennett *et al.* 1966).

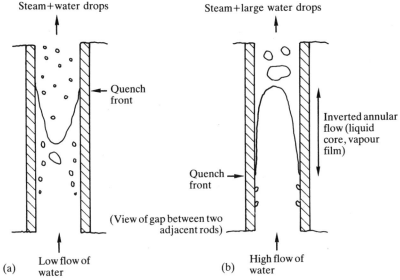

Fig. 21.7. Bottom flooding of a hot surface.

(a) The magnitude and variation of the heat transfer coefficient in the region of the quench front must be known.

(b) Heat conduction in the metal wall must be taken into account.

Taking a quench front advancing as shown in Fig. 21.8 and assuming, see Collier (1982)

(a) that the slab is infinite in the z direction,
(b) that it is insulated at $y = 0$,

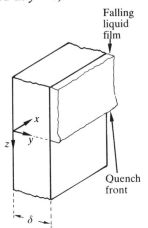

Fig. 21.8. Advance of a quenching front: co-ordinate system.

(c) that the liquid wets the surface at $z = 0$ when $T = T_0$ (a constant: the spluttering temperature), and

(d) that the position of the quench front is a function of z only at a given time, then the problem can be non-dimensionalized as follows.

The conduction equation is

$$\alpha \nabla^2 T = \frac{\partial T}{\partial t}. \tag{21.19}$$

The x variation is not important, and so

$$\frac{\partial^2 T}{\partial y^2} + \frac{\partial^2 T}{\partial z^2} = \frac{1}{\alpha} \frac{\partial T}{\partial t}. \tag{21.20}$$

In these equations α is the thermal diffusivity (m²/s), defined by

$$\alpha = k/\rho C_p, \tag{21.21}$$

where the physical properties k, ρ, and C_p are the properties of the tube wall material.

If the quench velocity u is constant, and the temperature variation as seen from the moving quench front is constant, then

$$\frac{\partial T}{\partial t} = -u \frac{\partial T}{\partial z}, \tag{21.22}$$

and so

$$\frac{\partial^2 T}{\partial y^2} + \frac{\partial^2 T}{\partial z^2} + \frac{u}{\alpha} \frac{\partial T}{\partial z} = 0. \tag{21.23}$$

Now, using non-dimensional variables

$$\eta = \frac{z}{\delta} \tag{21.24}$$

$$\zeta = \frac{y}{\delta} \tag{21.25}$$

$$\theta = \frac{T - T_{sat}}{T_0 - T_{sat}}, \tag{21.26}$$

where T_0 is the spluttering temperature and T_{sat} is the saturation temperature, then

$$\frac{\partial^2 T}{\partial y^2} = \frac{\partial^2 \theta}{\partial \zeta^2} \frac{T_0 - T_{sat}}{\delta^2}, \tag{21.27}$$

$$\frac{\partial^2 T}{\partial z^2} = \frac{\partial^2 \theta}{\partial \eta^2} \frac{T_0 - T_{sat}}{\delta^2}, \tag{21.28}$$

and

$$\frac{\partial T}{\partial z} = \frac{\partial \theta}{\partial \eta} \frac{T_0 - T_{sat}}{\delta}. \tag{21.29}$$

So, substituting these results into eqn (21.23),

$$\frac{\partial^2 \theta}{\partial \zeta^2} \frac{T_0 - T_{sat}}{\delta^2} + \frac{\partial^2 \theta}{\partial \eta^2} \frac{T_0 - T_{sat}}{\delta^2} + \frac{u}{\alpha} \frac{\partial \theta}{\partial \eta} \frac{T_0 - T_{sat}}{\delta} = 0 \tag{21.30}$$

or

$$\frac{\partial^2 \theta}{\partial \zeta^2} + \frac{\partial^2 \theta}{\partial \eta^2} + Pe \frac{\partial \theta}{\partial \eta} = 0, \tag{21.31}$$

where the Peclet number Pe is

$$Pe = u\delta/\alpha. \tag{21.32}$$

The four necessary boundary conditions are as follows, in terms of both the real and the dimensionless variables.

	Real variables		*Dimensionless variables*	
(a)	$z = -\infty,$	$T = T_{sat}$	$\eta = -\infty,$	$\theta = 0$
(b)	$z = +\infty,$	$T = T_w$	$\eta = +\infty,$	$\theta = \theta_w = \dfrac{T_w - T_{sat}}{T_0 - T_{sat}}$
(c)	$\dfrac{\partial T}{\partial y} = 0,$	$y = 0$	$\dfrac{\partial \theta}{\partial \zeta} = 0,$	$\zeta = 0$
(d)	$-k\dfrac{\partial T}{\partial y} = h(T - T_{sat}),$	$y = \delta$	$\dfrac{\partial \theta}{\partial \zeta} = -Bi\theta,$	$\zeta = 1$

T_w is the original wall temperature, and the Biot number Bi is

$$Bi = \frac{h\delta}{k}. \tag{21.32}$$

Hence, non-dimensionalizing the problem has shown that the important numbers are

(a) the Biot number $\dfrac{h\delta}{k}$,

(b) the Peclet number $\dfrac{u\delta}{\alpha}$, and

(c) the dimensionless wall temperature $\dfrac{T_w - T_{sat}}{T_0 - T_{sat}}$.

The solution of the problem is still far from easy for a number of reasons.

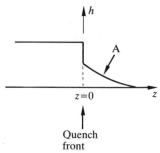

Fig. 21.9. Possible heat transfer coefficient variation in the neighbourhood of the quenching front (from Sun *et al.* 1975).

First, the non-dimensional equation is a second-order partial differential equation where two of the boundary conditions are themselves in the form of differential conditions. One simplifying assumption is to assume a particular form for the term $\partial^2 T/\partial y^2$ (or equivalently $\partial^2 \theta/\partial \zeta^2$), and so the equation becomes a second-order ordinary differential equation.

Secondly, the heat transfer coefficient h is not a constant but is a function of z. Many forms have been assumed; a typical form taken is shown in Fig. 21.9. The heat transfer in the region marked A ahead of the quench front is due to drops of water impinging on the hot surface ahead of the front. This phenomenon is known as *precursory cooling*.

Finally, some assumption, a numerical value, must be made about the spluttering temperature T_0. This is a real difficulty because, as has been noted earlier, the spluttering temperature is dependent upon the surface condition. Inevitably with the available choices of spluttering temperature and heat transfer coefficient variation (as illustrated in Fig. 21.9), a reasonable fit with experimental results can be obtained.

References

BENNETT, A. W., HEWITT, G. F., KEARSEY, H. A., and KEEYS, R. K. F. (1966). The wetting of hot surfaces by water in a steam environment at high pressure. *AERE-R5146.*

BERGLES, A. E. (1981). Two-phase flow and heat transfer. In *Two phase flow and heat transfer in the power and process industries* (A. E. Bergles, J. G. Collier, J. M. Delhaye, G. F. Hewitt, and F. Mayinger). McGraw-Hill/Hemisphere, Washington.

BERGLES, A. E. and THOMPSON, W. G. (1970). The relationship of quench data to steady state pool boiling data. *Int. J. Heat Mass Transfer* **13,** 55–68.

BOLLE, L. and MOUREAU, J. C. (1982). Spray cooling of hot surfaces. In *Multiphase science and technology* **1,** 1–97 (ed. G. F. Hewitt, J. M. Delhaye, and N. Zuber). McGraw-Hill/Hemisphere, Washington.

COLLIER, J. G. (1982). Heat transfer in the post-burnout region and during quenching and reflooding. In *Handbook of multiphase systems* (ed. G. Hetsroni). McGraw-Hill, New York.

SUN, K. H., DIX, G. E., and TIEN, C. L. (1975). The effect of precursory cooling on falling film rewetting. *J. Heat Transfer* **96**, 126–31.

ZUBER, N. and TRIBUS, M. (1958). Further remarks on the stability of boiling heat transfer. *Report 58-5*. Univ. of California, Los Angeles.

22

CONDENSATION

22.1 Modes of condensation

Condensation can occur in two ways, as illustrated in Fig. 22.1. First there could be filmwise condensation, as in Fig. 22.1(a), where the condensate (that is the liquid formed when the vapour condenses) forms a continuous film on the solid surface. The latent heat released on condensation is conducted through this film from the interface where the condensation occurs and is removed through the wall. The other mechanism is dropwise condensation, shown in Fig. 22.1(b). In this case the condensate forms in drops which do not wet the solid surface well. The drops therefore do not form a continuous film; instead they reach a certain critical size and then run off the surface. The surface is left dry and another drop can begin to form, rather like the formation of bubbles in nucleate boiling from a nucleation site. Because in dropwise condensation the vapour is in direct contact with the solid surface, the condensation heat transfer coefficients are significantly larger than in filmwise condensation. However, dropwise condensation is difficult to promote reliably. Special surface coatings do give drop formation, but the effectiveness of such surface treatments decreases with time. The normal design practice is to assume that filmwise condensation takes place.

For a general review of condensation see Griffith and Butterworth (1982).

22.2 Filmwise condensation on a vertical surface

For a laminar falling film the calculation of the heat transfer coefficient can be split into a number of parts.

(a) *Hydrodynamics.* As was done for flooding flow in § 11.2, consider a force balance on the outer part of the film (see Fig. 22.2). Now, however, the gas density is taken into account. For steady flow, in which the streamlines in the liquid are vertical

$$\delta p = \rho_g g \delta z, \tag{22.1}$$

and so a force balance in the vertical direction gives

$$\tau \delta z + \delta p(\delta - y) = \rho_l g(\delta - y)\delta z \tag{22.2}$$

or simplifying using eqn (22.1)

$$\tau = (\rho_l - \rho_g)g(\delta - y). \tag{22.3}$$

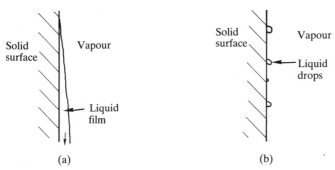

Fig. 22.1. Modes of condensation: (a) filmwise condensation, and (b) dropwise condensation.

Then, putting the laminar flow value for the shear stress

$$\tau = \mu_1 \frac{du}{dy} \qquad (22.4)$$

and so, using the boundary condition that $y = 0$, $u = 0$

$$u = \frac{(\rho_1 - \rho_g)gy}{\mu_1} \left(\delta - \frac{y}{2} \right). \qquad (22.5)$$

Note that we have also put (implicitly) that the interfacial shear stress is zero, that is that $\tau = 0$ when $y = \delta$. The mass flow rate M per unit width of film is then given by

$$M = \int_0^\delta \rho_1 u \, dy = \frac{\rho_1(\rho_1 - \rho_g)g}{\mu_1} \frac{\delta^3}{3} \qquad (22.6)$$

and differentiating, the change in the mass flow rate can be related to the

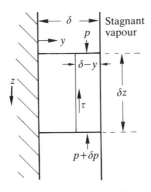

Fig. 22.2. Control volume for force balance in falling film flow.

changes in the film thickness

$$\frac{dM}{dz} = \frac{\rho_1(\rho_1 - \rho_g)g}{\mu_1} \delta^2 \, d\delta/dz. \tag{22.7}$$

Equation (22.7) is then valid as long as the film thickness changes slowly so that the streamlines are nearly vertical. This is true as long as $d\delta/dz \ll 1$.

(b) *Heat transfer.* As the vapour condenses its specific enthalpy changes by λ the latent heat. Strictly, the change is slightly larger since the condensate becomes subcooled in the liquid film. However, to a good approximation

$$\phi = \lambda \frac{dM}{dz}. \tag{22.8}$$

Another equation for the heat flux comes from the fact that in laminar flow the heat is conducted through the film: there is no convection. The temperature profile is linear, so from the conduction equation

$$\phi = +\frac{k_1(T_w - T_{sat})}{\delta} = +\frac{k_1}{\delta} \Delta T_{sat}, \tag{22.9}$$

where, in this equation, the positive sign arises because:

at $y = 0$, $T = T_w$, and
at $y = \delta$, $T = T_{sat}$.

Equating eqns (22.8) and (22.9) for the heat flux

$$\frac{dM}{dz} = \frac{k_1 \Delta T_{sat}}{\lambda \delta} \tag{22.10}$$

(c) *Solution for heat transfer coefficient.* Now we have two equations for dM/dz; one from the hydrodynamics (eqn (22.7)) and one from the heat transfer (eqn (22.10)). Equating the two values of dM/dz, and rearranging gives

$$\delta^3 \, d\delta = \frac{\mu_1 k_1 \Delta T_{sat}}{\rho_1(\rho_1 - \rho_g)g\lambda} \, dz. \tag{22.11}$$

Then integrating, and assuming that the film thickness is zero at $z = 0$, eqn (22.11) gives

$$\delta = \left[\frac{4\mu_1 k_1 \Delta T_{sat} z}{\rho_1(\rho_1 - \rho_g)g\lambda} \right]^{\frac{1}{4}}, \tag{22.12}$$

where δ is the film thickness at a distance z from the top of the plate. The heat transfer coefficient h_z at this distance z is then given by

$$h_z = k_l/\delta. \tag{22.13}$$

So, using eqn (22.12),

$$h_z = \left[\frac{\rho_l(\rho_l - \rho_g)g\lambda k_l^3}{4\mu_l \Delta T_{sat} z} \right]^{\frac{1}{4}}. \tag{22.14}$$

However, we are usually interested, not in the local heat transfer coefficient h_z, but the average value h over the whole plate. This value can be obtained by averaging h_z

$$h = \frac{1}{L} \int_0^L h_z \, dz. \tag{22.15}$$

The result of this integration is that the average condensation heat transfer coefficient h is given by

$$h = \frac{2\sqrt{2}}{3} \left[\frac{\rho_l(\rho_l - \rho_g)g\lambda k_l^3}{\mu_l \Delta T_{sat} L} \right]^{\frac{1}{4}}. \tag{22.16}$$

This equation was first obtained by Nusselt in 1916. The effect of the subcooling of the liquid in the film is easily taken into account if it is assumed that there is a linear temperature profile in the film. The result is that the latent heat in eqn (22.16) should be replaced by a modified latent heat λ'

$$\lambda' = \lambda + \tfrac{3}{8} C_{pl}(T_{sat} - T_w). \tag{22.17}$$

Rohsenow (1956) pointed out that because the condensation is occurring all the time, the film never has a chance to adopt the linear temperature profile. Because of this effect the factor of $\tfrac{3}{8}$ in eqn (22.17) should be modified to 0.68.

Example

Calculate the average heat transfer coefficient of condensation for steam at 1 bar on a vertical surface 1 m long when the surface temperature is 60 °C. The physical properties are:

$\rho_l = 1000 \text{ kg/m}^3$,
$\lambda = 2256 \times 10^3 \text{ J/kg}$,
$\rho_g \approx 0$,
$k_l = 0.68 \text{ W/m K}$, and
$\mu_l = 0.283 \times 10^{-3} \text{ N s/m}^2$.

The remaining variables required are then $\Delta T_{sat} = 100 - 60 = 40$ °C, and

$L = 1$ m, so from eqn (22.16)

$$h = \frac{8^{\frac{1}{2}}}{3}\left(\frac{10^3 \times 10^3 \times 9.81 \times 2256 \times 10^3 \times 0.68^3}{0.283 \times 10^{-3} \times 40 \times 1}\right)^{\frac{1}{4}} = 4700 \text{ W/m}^2 \text{ K}.$$

This is a typical value for the condensation heat transfer coefficient. A further example is given in Chaper 25 (see Example 11).

22.3 Filmwise condensation on a horizontal tube

For condensation on the outside of a horizontal tube, the condensate film drains down the tube and drips off the bottom as shown in Fig. 22.3. The derivation of the equation for the average condensation heat transfer coefficient is similar to, but more complicated than, the derivation of eqn (22.16). The final result is that for the horizontal tube

$$h = 0.73\left[\frac{\rho_l(\rho_l - \rho_g)g\lambda k_l^3}{\mu_l \Delta T_{\text{sat}} d}\right]^{\frac{1}{4}}. \tag{22.18}$$

The equation is in exactly the same form as eqn (22.16), except that the constant in the equation for a flat plate was $8^{\frac{1}{2}}/3 = 0.94$. The present value is lower because on a tube the film tends to be thicker. An application of this equation, coupled with heat transfer at the inside tube surface is given in Example 11 in Chapter 25. Typical heat transfer coefficients on a horizontal tube are larger than for a flat plate, so repeating the example in § 22.2 with $d = 0.02$ m instead of $L = 1$ m, gives $h = 9700 \text{ W/m}^2 \text{ K}$ (about twice the previous value). For this reason, condenser tubes are usually arranged to be horizontal rather than vertical. However, we usually have bundles of horizontal tubes not just one, as in Fig. 22.4. The condensate from one tube drips on to the next one below. The effect is almost as if the tube diameter was larger than the actual value. In the vertical plate derivation this would mean that the film thickness was not equal to zero at $z = 0$ for the second and subsequent plate in a vertical line. So, if there are n tubes in a vertical line, then the average condensation heat transfer coefficient h_n over the n

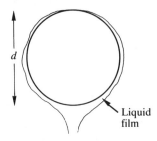

Fig. 22.3. Filmwise condensation on the outside of a horizontal tube.

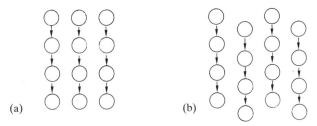

Fig. 22.4. Drainage of condensate in a tube bundle: (a) in-line array, and (b) staggered array.

tubes is given by

$$h_n = h/n^{\frac{1}{4}}. \tag{22.19}$$

So, if $h = 9700 \text{ W/m}^2 \text{ K}$ (as above), and $n = 18$, then

$$h_n = 9700/18^{\frac{1}{4}} = 4700 \text{ W/m}^2 \text{ K},$$

which is approximately the same result as for the vertical plate of length 1 m.

Actually, the downward flow pattern of the condensate is more favourable than envisaged in the simple extension to the Nusselt analysis, as illustrated in Fig. 22.5. Figure 22.5(a) shows the idealized flow, where the liquid flows off one tube in a continuous sheet. However, this is not what actually happens: as shown in Fig. 22.5(b), the liquid drips off at discrete points. The separation of the drip points is roughly given by the Taylor wavelength

$$\lambda_T = 2\pi\sqrt{3}\left[\frac{\sigma}{(\rho_1 - \rho_g)g}\right]^{\frac{1}{2}}.$$

as in § 15.2 on pool-boiling critical heat flux. This alteration to the ideal

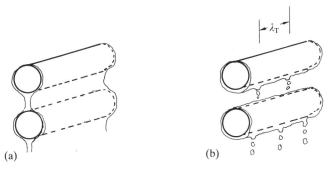

Fig. 22.5. (a) Idealized 'Nusselt' condensation and (b) actual flow of condensate from a horizontal tube to a tube below.

flow means that the $n^{-\frac{1}{4}}$ factor in eqn (22.19) is not correct. Experimentally, a better version of that equation is (see Collier 1981)

$$h_n = h/n^{\frac{1}{6}}. \tag{22.20}$$

So, taking the figures used before, $h = 9700 \text{ W/m}^2 \text{ K}$ and $n = 18$, then

$$h_n = 6000 \text{ W/m}^2 \text{ K}.$$

Real bundles are usually arranged:

(a) so that the tubes are staggered as in Fig. 22.4(b), not in a vertical line as in Fig. 22.4(a) (this means that the condensate from a tube at the top of the bundle flows over as few other tubes as possible); and

(b) so that in very large bundles the condensate is removed at intervals down the bundle (so that n in eqn (22.20) does not get too large).

22.4 Real condensation

Real condensation differs from the above analysis (the Nusselt analysis) in a number of ways.

(a) *The film is almost never smooth.* Ripples on the surface increase the surface area and stir up the film. These effects can increase the heat transfer coefficient by about 20 %.

(b) *There is a shear force exerted by the flowing vapour on the liquid film.* Normally, this is arranged to make the film thinner and so to increase the heat transfer coefficient.

(c) *The film may become turbulent.* This happens at values of the film Reynolds number $(4M/\mu_l)$ of 1000 to 2000. Again, turbulence increases the heat transfer coefficient as compared to the laminar flow values. The effect of the turbulence is to increase the effective viscosity of the liquid: this makes the liquid film thicker. This effect is, however, more than counteracted by the increased heat transfer due to convection effects in the liquid.

(d) *Multi-component effects.* The simplest of these is the effect of an 'incondensable' gas (such as, for example, air) on the condensation of steam. Such effects are discussed in some detail in § 22.9. However, it can be noted here that in many cases the presence of an incondensable gas can give rise to condensation heat transfer coefficients which are far less than expected.

In contrast to the large effects of an incondensable gas, the first three effects all lead to fairly modest increases in the condensation heat transfer coefficient.

22.5 Condensation inside a horizontal tube

Condensation inside a horizontal tube is a common industrial situation. The vapour may enter the tube superheated, so the flow will be as shown in Fig. 22.6. At point A the wall is first wet with condensate, the liquid

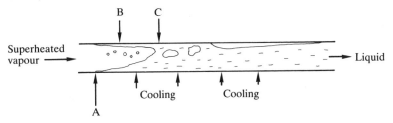

Fig. 22.6. Total condensation of a pure vapour in a horizontal tube.

film on the walls of the tube is slightly subcooled and the vapour is still superheated. The region before A is known as the dry wall desuperheating region; the region after A is known as the wet wall desuperheating region. At point B liquid drops are torn off the liquid film into the superheated vapour where they then evaporate. From A until C the flow is an annular flow. At first, near A, there is a significant degree of thermodynamic non-equilibrium (superheated vapour and subcooled

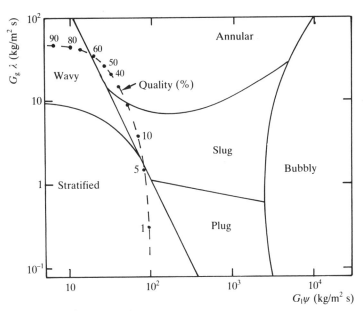

Fig. 22.7. Condensation path superimposed on the Baker flow pattern chart.

liquid). It can be seen that the region from A to C is a possible region for the application of annular flow modelling as described in § 20.6 on post-dryout heat transfer. This has indeed been done but, as in post-dryout heat transfer, the work has been hampered by a lack of detailed experimental data for anything except wall temperature. From point C onwards the flow goes through various patterns until the last vapour condenses (in a total condenser) to give, finally, subcooled liquid. In analysing the condensation, it is interesting to note the flow patterns which are found. The path of the condensation can be plotted on a flow regime map; in Fig. 22.7 a typical condensation path is plotted on a Baker flow pattern map (see § 2.6), where λ and ψ are defined. As in boiling flow, a substantial length of the tube is in the annular flow region. The condensation coefficient has been analysed for various flow patterns. This work is described in the following sections; for further details see Collier (1981) and Owen and Lee (1983).

22.6 Stratified flow

In general the quality here is low and so there is a low interfacial shear stress. The Nusselt analysis can hence be legitimately used, so that

$$h = F\left[\frac{\rho_l(\rho_l - \rho_g)g\lambda k_l^3}{\mu_l \Delta T_{sat}d}\right]^{\frac{1}{4}}. \tag{22.21}$$

For condensation outside a horizontal tube, the numerical factor F is equal to 0.73. For condensation inside a tube in stratified flow (see Fig. 22.8), it is commonly assumed that no condensation occurs in the stratified layer and that all the condensation occurs in the top part of the tube. At the top of the tube the flow is very similar to the flow in

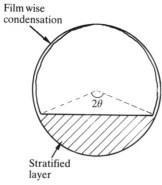

Fig. 22.8. Stratified flow.

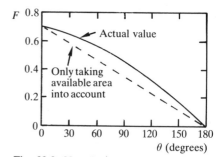

Fig. 22.9. Nusselt equation correction factor for stratified flow.

condensation outside a horizontal tube. The value of F changes with the angle θ (see Fig. 22.8) because of the following.

(a) Less of the tube wall area is available for condensation. Taking the area effect crudely into account would suggest that F declines linearly as θ increases.

(b) At the top the tube is relatively thin. This means that the top of the tube is relatively more efficient in condensation terms than the bottom, so the fall-off in the value of F is not linear, as illustrated in Fig. 22.9.

Sometimes, in the absence of other information, a value of $F = 0.58$, which is equivalent to a value for θ of 60°, is taken.

22.7 Slug and plug flow

In this case there is some condensation near the bottom of the tube, and there are effects of vapour shear on the film at the top of the tube, so the position is quite complicated. Typical experimental results therefore show a variation in the heat transfer coefficient around the tube, as shown in Fig. 22.10. The plateau at the top of the tube can be predicted reasonably from

$$F = 0.31 Re_g^{0.12}, \tag{22.22}$$

where

$$Re_g = \frac{G_g d}{\mu_g} \tag{22.23}$$

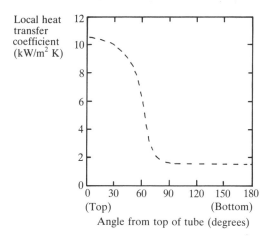

Fig. 22.10. Variation of condensation heat transfer coefficient around the tube periphery in slug flow.

22.8 Annular flow

In annular flow two approaches are possible.

(*a*) Using the universal velocity profile (as for boiling in annular flow), the heat transfer coefficient can be obtained as a function of the physical properties and the dimensionless film thickness. The method is exactly the same as for boiling flow, and is given in detail in § 19.3.

(*b*) A much simpler approach is to consider a film flow as in Fig. 22.11, and to work out the Reynolds number. If the volume flow rate of liquid (m^3/s) is denoted by Q_1, and mean film thickness is δ, then the area occupied by the film is approximately $\pi d\delta$ and the mean velocity of the liquid in the film is $Q_1/\pi d\delta$. Using a length scale for the Reynolds number of 4δ, then

$$Re = \frac{\rho_1 4\delta}{\mu_1}\frac{Q_1}{\pi d\delta} = \frac{4\rho_1}{\mu_1}\frac{Q_1}{\pi d}. \qquad (22.24)$$

For the same liquid volume flow rate occupying the whole tube in single-phase flow, then the Reynolds number can again be evaluated with

$$\text{mean velocity} = \frac{4Q_1}{\pi d^2},$$

$$\text{length scale} = d,$$

$$Re = \frac{\rho_1}{\mu_1}\frac{d4Q_1}{\pi d^2} = \frac{4\rho_1}{\mu_1}\frac{Q_1}{\pi d}. \qquad (22.25)$$

Note, the Reynolds numbers for the two flows are equal; they have been made equal by the appropriate choice of length scale (4δ) for the film flow case. So now we can write for both the above flows

$$Nu = f(Re, Pr), \qquad (22.26)$$

and for both the flows the Reynolds numbers are equal and the Prandtl numbers are equal. Then, making the assumption that the functional relationship in eqn (22.26) is the same in both cases, the Nusselt numbers

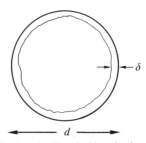

Fig. 22.11. Annular flow inside a horizontal tube.

in the two cases must be the same. Now, in general,

$$Nu = \frac{hl}{k_1},$$ (22.27)

where l is the appropriate length scale. Therefore

$$\frac{h}{h_1} = \frac{d}{4\delta} \simeq \frac{1}{1 - \alpha},$$ (22.28)

where h is the condensation heat transfer coefficient in the film flow (W/m^2 K), and h_1 is the single-phase liquid heat transfer coefficient (W/m^2 K) calculated using the liquid flow rate in the film flow.

Using either of these methods, the condensing heat transfer coefficient seems to be better calculated in annular flow than the corresponding boiling heat transfer coefficient. The reason for this is not clear.

22.9 Condensation of a vapour and an incondensable gas

In a case such as the condensation of steam which contains some air, there are concentration gradients of the air and the steam near the interface as shown in Fig. 22.12.

Although in this case the partial pressure of the air in the condenser is quite small, the partial pressure at the interface is quite substantial. The air is diffusing away from the interface down its concentration gradient, and being pulled towards the interface by the flow of steam. The air partial pressure profile is the result of these competing effects. It is evident that a proper calculation of the condensation rate would involve a complex boundary layer calculation. Such calculations are only now beginning to be done. Practical calculation of condensation coefficients

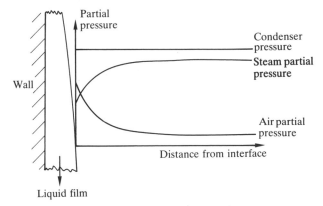

Fig. 22.12. Partial pressure of air and steam near the interface in a condenser.

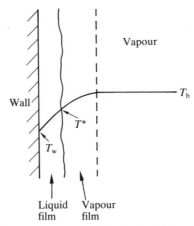

Fig. 22.13. Variation of temperature near and in the liquid film during the condensation of
a vapour containing an incondensable gas.

has relied on a number of drastically simplified theories, such as that of
Silver (1947) and Bell and Ghaly (1973). Plotting the temperature
distribution near the wall (see Fig. 22.13), the whole temperature
difference, $T_b - T_w$ can be divided into two parts

$$T_b - T_w = (T_b - T^*) + (T^* - T_w). \tag{22.29}$$

Assuming the heat flux is the same across the interface as across the wall
then, since $\Delta T = \phi/h$, eqn (22.29) can be rewritten as

$$\frac{\phi_w}{h} = \frac{\phi_g}{h_g} + \frac{\phi_w}{h_f}, \tag{22.30}$$

where ϕ_w is the wall heat flux (W/m^2), ϕ_g is the 'sensible' heat flux
(W/m^2)—this is the heat flux which arises because of the temperature
differences in the gas phase, h is the condensation heat transfer
coefficient $(W/m^2 K)$, h_f is the heat transfer coefficient across the liquid
film $(W/m^2 K)$ (this could be calculated by means of eqn (22.16)), and h_g
is the single-phase gas heat transfer coefficient $(W/m^2 K)$.

The energy flow from the vapour core to the interface is made up of
two parts:

(a) that which is released as latent heat by condensation at the
interface, and
(b) that which is represented by cooling of the vapour as it approaches
the interface: this is the 'sensible' heat.

It is assumed that the sensible heat resistance is at least as great as the
liquid film resistance.

We can now write

$$\frac{1}{h} = \frac{1}{h_f} + \frac{\phi_g/\phi_w}{h_g}.$$ (22.30)

In order to produce a solution, we need a value for ϕ_g/ϕ_w, which can be estimated by arguing that

$$\phi_g = \dot{m} C_{pg} \Delta Tx$$ (22.31)

where $\dot{m}$ is the total mass flow towards the surface per unit area of wall (kg/m^2 s), and x is the quality of this mass flow. Then the total heat flux ϕ_w is given by

$$\phi_w = \dot{m} \Delta h$$ (22.32)

Eliminating $\dot{m}$ between eqns (22.31) and (22.32) we find that

$$\frac{\phi_g}{\phi_w} = x C_{pg} \frac{dT}{dh},$$ (22.33)

where x is the local quality, C_{pg} is the gas-phase specific heat, and dT/dh is the slope of the condensation curve. The condensation curve is a graph of the enthalpy released against temperature as the condensation proceeds (see Fig. 22.14). The enthalpy is the total enthalpy at any point: the sum of the enthalpies of the liquid phase, of the condensable vapour, and of the non-condensable gas.

As an example, consider the calculation of a mixture originally of:

steam flow rate   0.9 kg/s,
air flow rate   0.1 kg/s, and at
total pressure p 1 bar.

The initial partial pressures are calculated from the ratio of the numbers of moles flowing

$$p_{steam} = \frac{\dfrac{0.9}{18}}{\dfrac{0.9}{18} + \dfrac{0.1}{29}} = 0.935 \text{ bar.}$$

$$p_{air} = p - p_{steam} = 1 - 0.935 = 0.065 \text{ bar.}$$

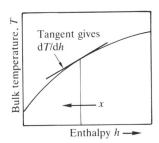

Fig. 22.14. Condensation curve.

Here 18 and 29 are the molecular weights of steam and air. The condensation will begin when the temperature is the steam saturation temperature equivalent to 0.935 bar, which is 97.6 °C. Then, choosing temperatures below this, the condensation curve can be constructed. For example, take $T = 90$ °C. At this temperature

$$p_{steam} = p_{sat} = 0.701 \text{ bar},$$

and so

$$p_{air} = p - p_{steam} = 1 - 0.701 = 0.299 \text{ bar}.$$

The volume occupied by 0.1 kg of air (the mass flowing in 1 s) at this temperature $(273 + 90 = 363 \text{ K})$ and pressure $(0.299 \times 10^5 \text{ N/m}^2)$ can be found from

$$V = \frac{m_{air}RT}{p_{air}} = \frac{0.1 \times \dfrac{8314}{29} \times 363}{0.299 \times 10^5} = 0.348 \text{ m}^3.$$

From steam tables, the specific volume of saturated steam at 90 °C is 2.361 m^3/kg, so the mass of steam associated with 0.1 kg of air is

$$\frac{0.348}{2.361} = 0.147 \text{ kg}.$$

Hence $0.9 - 0.147 = 0.753$ kg of steam have condensed. The enthalpy of the mass flowing in 1 s can now be calculated

$$h = 0.1 \times 1.005 \times 90 + 0.753 \times 376.9 + 0.147 \times 2660.1 = 683.9 \text{ kJ/kg},$$
$$\quad m_{air} \quad C_{p,air} \quad T \quad m_{water} \quad h_{water} \quad m_{steam} \quad h_{steam}$$

and

$$x = \frac{0.1 + 0.147}{1} = 0.247.$$

Note that just before condensation began, $T = 97.6$ °C, $x = 1$, and

$$h = 0.1 \times 1.005 \times 97.6 + 0.9 \times 2669 = 2411.9 \text{ kJ/kg}.$$
$$\quad m_{air} \quad C_{p,air} \quad T \quad m_{steam} \; h_{steam}$$

Taking other temperatures, the condensation curve can be calculated: the result is shown in Fig. 22.15.

Now since, from eqns (22.30) and (22.33)

$$\frac{1}{h} = \frac{1}{h_f} + \frac{xC_{pg} \, dT/dh}{h_g}, \tag{22.34}$$

h falls rapidly as the quality falls. This is because dT/dh becomes very large and so $1/h$ is large and therefore h is small. This means that the first

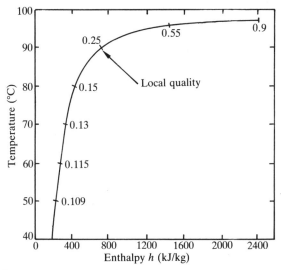

Fig. 22.15. Condensation curve for air–steam mixture used in the example in § 22.9.

part of the condensation can be performed without difficulty, but that it is very difficult (that is, a large surface area is required) to condense a very large fraction of the steam. An example (Example 11) using this condensation curve to calculate heat transfer coefficients is given in Chapter 25. Of course, the shape of the condensation curve alters as the inlet flows of air and steam are altered (or more specifically, as their ratio is altered). Reduction of the overall heat transfer coefficient is not so severe if the initial fraction of air in the steam, here assumed to be 10 % by weight, is lower.

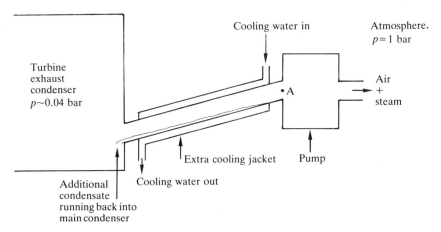

Fig. 22.16. Venting arrangement in a turbine exhaust condenser.

However, turbine exhaust condensers in a steam power station always have pumps to remove the air which inevitably leaks into the system. They are arranged so as to remove as much air and as little steam as possible (as the steam lost represents a loss of special water which is chemically treated in order to inhibit corrosion, and so as to minimize the pump power). One arrangement is shown in Fig. 22.16.

The object is to reduce the steam partial pressure at A to as low a value as possible. This is dependent on the temperature of the cooling water: obviously the lower the temperature of the cooling water the better.

References

BELL, K. J. and GHALY, M. A. (1973). An approximate generalised design method for multi-component/partial condensers. *AIChE Symp. Ser.* **69**, 72–9.

COLLIER, J. G. (1981). *Convective boiling and condensation* 2nd edn, McGraw-Hill, New York.

GRIFFITH, P. and BUTTERWORTH, D. (1982). Condensation. In *Handbook of multiphase systems* (ed G. Hetsroni). McGraw-Hill, New York.

OWEN, R. G. and LEE, W. C. (1983). Some recent developments in condensation theory. *Chem. Eng. Res. Des.* **61**, 335–61.

ROHSENOW, W. M. (1956). Heat transfer and temperature distribution in laminar film condensation. *Trans. ASME* **78**, 1645–8.

SILVER, L. (1947). Gas cooling with aqueous condensation. *Trans. Instn. Chem. Engnrs* **25**, 30–42.

23

PROCESS INDUSTRY REBOILERS

23.1 Introduction

Some of the more practical aspects of boiling are considered here. The examples are taken from the process industry. Of the boiling operations in the process industry, one of the most common is that to provide the vapour for a distillation column. In a distillation column vapour moves up the column and is brought into contact with liquid which moves down. The vapour is provided by a reboiler which vaporizes most the liquid which arrives at the bottom of the column.

23.2 Types of reboiler

The common types of reboiler are all based on shell and tube heat exchangers; a simple shell and tube heat exchanger is shown in Fig. 23.1. One fluid flows along the tubes (the tube-side fluid) and the other flows outside the tubes and is contained in the shell (the shell-side fluid). The shell may, or may not, have baffles to direct the flow of the shell-side fluid. Various possibilities for a reboiler then arise: the process fluid (that is the fluid to be vaporized) can be boiled inside the tubes or outside, and the process fluid can be separated into liquid and vapour steams in the reboiler itself, or in the distillation column. Three types of reboiler are commonly in use.

(a) *Kettle reboiler* (*see Fig. 23.2*). Here the process fluid from the distillation column is vaporized on the shell side of the heat exchanger, and the separation of the vapour and the liquid is performed in the reboiler shell. In order to provide space for the disengagement of the liquid drops thrown up by the boiling from the vapour, a much enlarged shell is necessary. Commonly, the tube bundle only occupies the lower half of the shell cross section.

The advantages of the kettle reboiler are that the design is simple and control problems are minimized, but the heat exchanger itself is large and therefore expensive. This is a particular disadvantage at high pressure, as the shell wall thickness must be large. Kettle reboilers also suffer from fouling from the process liquid outside the tubes, and the outside of the tube is difficult to clean. The kettle reboiler is easy to design because as a

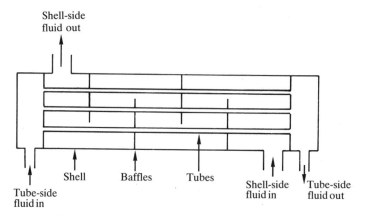

Fig. 23.1. Shell and tube heat exchangers.

first approximation:

    (i) the hydrodynamics of the system can be neglected; and
    (ii) the boiling process can be regarded as pure pool boiling.

It will be seen later (in § 23.3) that these assumptions are such that the heat transfer coefficient is underestimated and therefore the reboiler is overdesigned.

    (*b*) *Horizontal thermosyphon reboiler* (*see Fig.* 23.3). In a horizontal thermosyphon reboiler, the process fluid again is vaporized on the shell side. However, the reboiler returns a mixture of liquid and vapour to the column inside which it is separated.

    The advantages of a horizontal thermosyphon reboiler are that the shell is now much smaller in diameter and the plant layout is convenient,

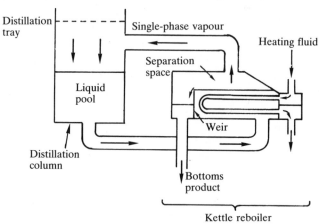

Fig. 23.2. Kettle reboiler.

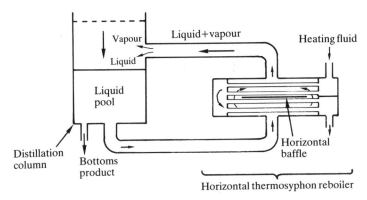

Fig. 23.3. Horizontal thermosyphon reboiler.

with the reboiler next to the distillation column. The disadvantages are that design is now difficult and there may be problems in operation: the reboiler may suffer from a kind of instability of the oscillatory nature described in Chapter 12. Oscillatory instability in boiling two-phase flow is a very complex phenomenon, but two things are clear (see Chapter 12).

(*i*) the pressure drop at the inlet to the boiling section is a stabilizing factor, and

(*ii*) the pressure drop at the outlet of the boiling section is a destabilizing factor.

The design is complicated because the flow rate through the reboiler and, to a lesser, though significant, extent, the heat transfer rate is determined by a natural circulation flow driven by density differences, mainly in the two-phase mixture.

(*c*) *Vertical thermosyphon reboiler* (*see Fig.* 23.4). In this system the fluid is vaporized on the tube side of the reboiler and the residence time in the reboiler is very short. This makes a vertical thermosyphon reboiler particularly suitable for fouling liquids and for heat sensitive liquids. The disadvantages are again that the reboiler is difficult to design (because of the thermosyphon action) and there may be stability problems. A further disadvantage is that the distillation column may have to be lifted up to accommodate the reboiler at the column base; however, compared with a horizontal thermosyphon reboiler only a small site area is required. The liquid level in the distillation column is commonly maintained at the top tube plate level of the reboiler. However, when the process fluid is at low pressure (less than 1 bar) a significant length of the tubes is used to heat the fluid up to its saturation temperature. So, for such low-pressure conditions, the liquid level in the distillation column is often reduced.

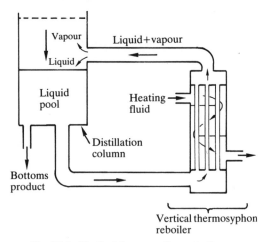

Fig. 23.4. Vertical thermosyphon reboiler.

The choice of reboiler type is often not simple (see Jacobs 1961; and Whalley and Hewitt 1986).

23.3 Calculation of flow rate in a thermosyphon

It is obvious that the horizontal and vertical thermosyphons depend on density differences to drive the circulation of the process fluid through the reboiler. This is also true, but less obvious, for the kettle reboiler. Looking at a cross section through the large diameter shell, the observed flow is shown in Fig. 23.5. There is a strong circulation of fluid up through the bundle, then the liquid returns to the bottom of the bundle

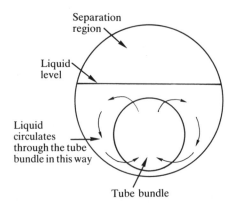

Fig. 23.5. Cross section of a kettle reboiler (from Brisbane *et al.* 1980).

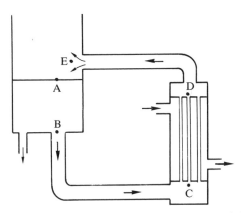

Fig. 23.6. Flow circuit in a vertical thermosyphon reboiler.

along the sides of the bundle. The flow up through the bundle is predominantly vertical. The detailed heat transfer coefficients depend on the flow circulation. However, the circulation is often ignored and boiling is therefore assumed to be pool boiling. This can lead to errors because the flow through the bundle may be very substantial (see Leong and Cornwell 1979).

Taking a vertical thermosyphon reboiler, in which the fluid flow path is more obvious, the process fluid flows around the cycle ABCDEA shown in Fig. 23.6. If all the pressure drops (some are actually pressure rises) around this circuit are known, the sum should be zero. The way the flow rate is calculated is to guess a flow rate and then to adjust it, often by trial and error, finally making the total pressure change around the circuit ABCDEA equal to zero. Taking the parts of the circuit in turn we get

A → B is essentially stagnant liquid, so there is only the hydrostatic head term.

B → C is single-phase flow in a pipe.

C → D is the most complicated part of the circuit. The liquid starts off subcooled at C because it was saturated at A, but the increase in pressure from A to C has increased the saturation temperature. So first, in the flow from C to D the liquid is heated to just above its local saturation temperature (which is changing as the local pressure changes) and boiling then begins. Now, the temperature of the fluid drops again as the pressure falls in going up the tube. In a well-designed thermosyphon the quality does not reach that at which critical heat flux occurs. Hence the temperature profile up the tube is as shown in Fig. 23.7.

D → E is a two-phase flow in a complex geometrical arrangement: from the tubes into the header, from the header into the outlet nozzle, and then along the outlet pipe. This involves geometries which have not been

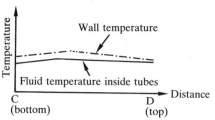

Fig. 23.7. Variation of wall and fluid temperatures in a vertical thermosyphon reboiler.

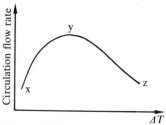

Fig. 23.8. Variation of circulation flow rate with temperature difference in a vertical thermosyphon reboiler (from Johnson 1956).

studied in detail, and so homogeneous flow is usually assumed. Note that as the pressure falls, some of the liquid phase will flash to form additional vapour.

E→A the pressure rises due to the small 'hydrostatic' head in the vapour.

As the heat flux, or the equivalent average temperature difference ΔT between hot and cold streams in the reboiler, is varied the circulation flow rate varies as shown in Fig. 23.8.

This behaviour occurs because there are two competing effects:

(a) as ΔT increases, so also does the void fraction in the tubes and so the mean density difference increases; and

(b) as ΔT increases, so also does the frictional two phase pressure drop.

From X→Y in Fig. 23.8, the first effect predominates and so the flow rate increases as ΔT increases, but the second effect becomes important from Y→Z and so in this latter region the circulation flow rate falls. There is obviously some merit in operating an actual reboiler near the peak in Fig. 23.8, as this will give the maximum convective effect on the heat transfer coefficient.

The temperature difference cannot be too high because at high values of ΔT dryout and film boiling will occur. Dryout is not usually dangerous in a fluid heated reboiler, but the decreased heat transfer coefficient means that the reboiler will not reach its design duty.

References

BRISBANE, T. W. C., GRANT, I. D. R., and WHALLEY, P. B. (1980). A prediction method for kettle reboiler performance. *ASME paper 80-HT-42.* (ASME/AIChE Heat Transfer Conference, Orlando, Florida).

JACOBS, J. K. (1961). Reboiler selection simplified. *Hydrocarbon Processing and Petroleum Refiner* **40** (No. 7), 189–96.

JOHNSON, A. I. (1956). Circulation rates and overall temperature driving forces in a vertical thermosyphon reboiler. *Chem. Eng. Prog. Symp. Ser.* **52** (No. 18), 37–46.

LEONG, L. S. and CORNWELL, K. (1979). Heat transfer coefficient in a reboiler tube bundle' *The Chemical Engineer* **343**, 219–21.

WHALLEY, P. B. and HEWITT, G. F. (1986). Reboilers. Published In *Multiphase science and technology* (ed. G. F. Hewitt, J. M. Delhaye, and N. Zuber) Vol. 2, pp. 275–331. Hemisphere Publishing Corporation, New York.

24

CONDENSERS

24.1 Introduction

There are many different types of condensers, (see Butterworth 1982 for details), but only two main types are considered here:

(*a*) process industry shell and tube condensers which can be divided into tube-side and shell-side condensers; and

(*b*) power industry turbine exhaust condensers which are often called surface condensers, but which are a specialized form of shell and tube heat exchanger.

Other common types of condenser are air-cooled condensers, plate condensers, and direct contact condensers.

In process industry shell and tube heat exchangers the choice between a tube-side condenser and a shell-side condenser is mainly based on a number of criteria.

(*a*) The most fouling fluid is normally handled on the tube side rather than on the shell side. This is because the inside of the tubes is easier to clean than the outside. A similar criterion is that the most corrosive fluid is best inside the tube because then the shell need not be made of special material.

(*b*) The higher pressure fluid is normally on the tube side. This is because the shell will then be less expensive to manufacture.

(*c*) Shell side condensers can be designed to have a lower pressure drop for the condensing fluid than can tube side condensers. If the pressure drop is an important factor, this may determine the type of condenser to be used.

This last criterion is important in turbine exhaust condensers where a low condensing side pressure drop contributes significantly to the thermodynamic efficiency of the whole cycle. Thus turbine exhaust condensers are invariably shell-side condensers, though of a specialized design.

One of the most important features of any condenser is that it should be equipped with a vent to remove, or at least to reduce, the concentration of any incondensable gas such as air.

24.2 Tube-side shell and tube condensers

Tube-side condensers are usually mounted with the shell nominally horizontal, although actually there is considerable advantage in having a small angle of inclination such that the condensate tends to run out of the end of the tubes at the opposite end to the vapour inlet (see Fig. 24.1(a)). This will make the flow much more stable than if the tubes are horizontal or inclined slightly upwards (see Fig. 24.1(b)). It also makes the venting system work more efficiently (the vent is placed at the tube outlet header).

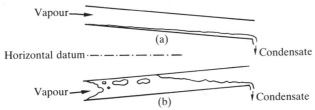

Fig. 24.1. Horizontal tube-side condensers, effect of slight inclinations of the exchangers: (a) downward sloping, and (b) upward sloping.

Normally, only a single tube-side pass is used, although two tube-side passes can be used if the first pass is inclined slightly upward. In this case the vapour velocity is sufficient, if the condenser has not been grossly overdesigned, to carry the liquid up this tube. The second pass is then inclined slightly downward so that the condensate liquid can run out of the tube, as in Fig. 24.1(a).

Multi-tube pass arrangements can lead to problems in distributing the flow at the inter-pass headers (see Fig. 24.2). A bottom tube (such as A in Fig. 24.2) tends to run full of liquid and add nothing to the heat load. An upper tube (such as B) tends initially to run with no liquid. These problems occur because the liquid and vapour from the first pass tend to separate in the inter-tube header, as shown in Fig. 24.2.

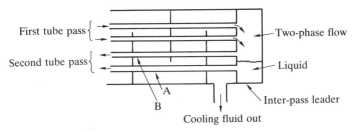

Fig. 24.2. Phase separation in a multi-tube pass tube-side condenser.

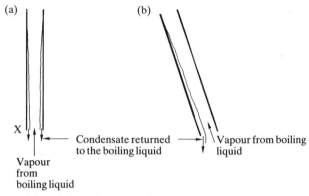

Fig. 24.3. Reflux tube-side condensers: (a) vertical and (b) inclined.

Tube-side condensers can also be mounted vertically. In this case the normal arrangement is for the condensing vapour to flow downwards so that again the condensate is able to drain out of the tubes easily. If the condensing vapour were flowing upwards the flow would be very unstable.

A special case is that of reflux condensation, in which the liquid condensate is returned to the boiling fluid. Commonly, the vapour flows upwards and the condensate drains back against the vapour flow (see Fig. 24.3(a)). Clearly, the liquid and the vapour flow rates in the tube are greatest at the bottom end of the tube (X in Fig. 24.3(a)). As there is counter-current flow, there is a danger that flooding will occur (see Chapter 11), and if flooding occurs anywhere it will occur at X. The condenser can be made less susceptible to flooding by inclining the tubes as in Fig. 24.3(b). Experiments on flooding have shown that it is less likely to occur when the tubes are inclined. This is because the liquid then flows down just part of the tube wall and the liquid and the vapour streams are thus better separated. However, designers are not always amenable to the idea of having an untidy-looking inclined condenser.

24.3 Shell-side shell and tube condensers

Shell-side condensers can be divided into baffled units and cross-flow (or unbaffled) units. A typical baffled unit, actually a TEMA (1978) E-type unit is shown in Fig. 24.4. The baffles are arranged so that the single segmented baffle cut is vertical and the flow is thus 'side-to-side' in the shell. If the baffle cut were horizontal, the flow would be 'up-and-down', giving a very high shell-side pressure drop and a rather unstable flow. The baffles (see Fig. 24.5) are often equipped with a notch (Fig. 24.5(a))

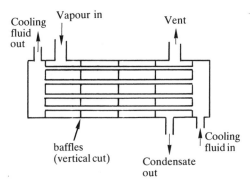

Fig. 24.4. Shell-side condenser (E-type shell).

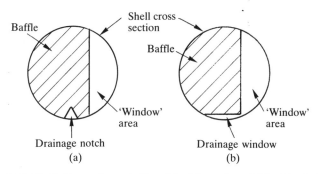

Fig. 24.5. Shell-side condenser baffles: (a) with notch, and (b) with window.

or additional small window (Fig. 24.5(b)) at the bottom of the shell to allow the condensate to drain along the shell more easily.

A typical cross-flow unit, a TEMA (1978) X-type unit, is shown in Fig. 24.6. There are a number of vapour inlet nozzles and a perforated distributor plate to help produce a pure cross flow over the tube bank, and so utilize the entire heat transfer area in the bundle. The 'baffles'

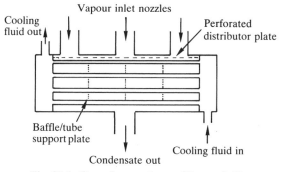

Fig. 24.6. Cross-flow condenser (X-type shell).

only serve to support the tubes and not to direct the flow. Cross-flow condensers are used when the overriding consideration is that the condensing side pressure drop should be as low as possible. The vent in a cross-flow unit is usually placed at the side of the shell, part way down the tube bank.

24.4 Turbine exhaust condensers

The particular characteristics of turbine exhaust condensers is that they condense steam at low pressure (approximately 0.04 bar) and there is only a limited temperature driving force (typically about 10 °C between the steam saturation temperature and the cooling water temperature) available. To condense the steam at as low a pressure as possible, the pressure drop in the condenser must be very small. The designs of turbine exhaust condensers are very specialized, and many different designs are used (see Editors of Power 1967; Simpson 1969). Butterworth (1982) gives a schematic diagram of a composite design (see Fig. 24.7). Note in this figure the following features.

(*a*) The shell is circular, but larger turbine exhaust condensers sometimes have rectangular shells.

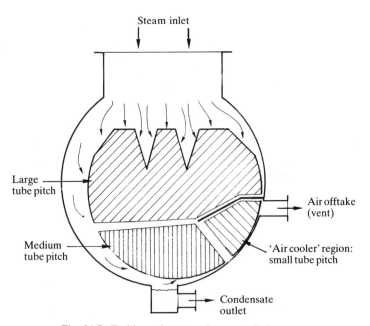

Fig. 24.7. Turbine exhaust condenser: main features.

(b) The steam inlet is of very large diameter, as the steam density is very low.

(c) The bundle contains steam access lanes, so that the steam can easily reach the centre of the bundle.

(d) The bundle contains tubes at different pitches. The steam first meets bundles at a large pitch, then the tube pitch becomes progressively smaller.

(e) The air vent is located at the side of the shell. Steam is prevented from entering the vent directly from the steam inlet by means of an internal baffle. This baffle forces the steam to flow over the part of the bundle with smallest tube pitch before reaching the vent. This part of the bundle is also cooled by the coldest water, so that the gases leaving the vent contain as much air and as little steam as possible. This reduces the steam loss and minimizes the compression work in raising the vent gas pressure to atmospheric pressure.

References

BUTTERWORTH, D. (1982). *Condensing equipment*. In *Handbook of multiphase systems* (ed. G. Hetsroni). McGraw-Hill, New York.

EDITORS OF POWER (1967). *Power generation systems,* pp. 282–9. McGraw-Hill, New York.

SIMPSON, H. C. (1969). Outline of current problems in condenser design. *Proc. Symp. to celebrate the Bicentenary of the James Watt Patent, Univ. of Glasgow,* pp. 131–4.

TEMA (1978). *Standards of the Tubular Exchanger Manufacturers Association* (6th edn). New York.

25

PRACTICAL CALCULATION
METHODS

Introduction

This chapter contains full details of a number of the calculation methods
described in the text which are likely to be of direct, practical interest.
The calculation methods which are included are listed below. For each
the relevant section, equation, and figure numbers are given. The
relevant text should be read carefully before the method is used to check,
for example, the range of applicability. Four of the more complicated
methods have been detailed separately in appendices. In every case the
SI system of units has been used.

Example	*Section*	*Eqns*	*Figs.*
Flow patterns in round tubes			
1(a). vertical flow	2.4		2.7
1(b). horizontal flow	2.4		2.8
Rising velocity of gas bubbles in stagnant liquids			
2. small spherical bubbles	3.1	3.3	
2. spherical cap bubbles	3.2	3.15	
2. intermediate size bubbles	3.2	3.6–8	3.3
3. gas plugs in vertical tubes	3.3	3.16	3.8
Homogeneous two-phase flow			
4(a). void fraction	5.1	5.5	
4(b). frictional pressure gradient	5.2–3	5.8,9,13	
Separated two-phase flow			
4(c). void fraction	6.5	see Appendix A	
4(d). frictional pressure gradient	6.9	see Appendix B	
Critical two-phase flow			
5(a)–(b). homogeneous frozen flow	8.3	8.20,22	
5(c). homogeneous equilibrium flow (for steam–water only)	8.5		8.3
Flooding in vertical tubes			
6. flooding velocities	11.7	11.18–20	

EXAMPLE 1 FLOW PATTERNS 243

Example 1 Flow patterns

Find the flow pattern when 4 kg/s of steam–water mixture of quality 20 % at 20 bar flows in a circular tube of internal diameter 0.1 m:

(a) when the flow is vertically upwards; and

(b) when the flow is horizontal.

Solution

The physical properties required are:

water density $\rho_l = 850 \text{ kg/m}^3$;
steam density $\rho_g = 10 \text{ kg/m}^3$;
water viscosity $\mu_l = 128 \times 10^{-6} \text{ N s/m}^2$; and
steam viscosity $\mu_g = 16 \times 10^{-6} \text{ N s/m}^2$.

(a) *Vertical upflow*. The method used is the flow pattern map of

Hewitt and Roberts (see Fig. 2.7). The axes are G_g^2/ρ_g and G_l^2/ρ_l. The total mass flux G is given by

$$G = \frac{\text{mass flow rate}}{\text{cross sectional area}} = \frac{4}{\pi 0.1^2/4} = 509 \text{ kg/m}^2 \text{ s}.$$

The phase mass fluxes are then

$$G_g = xG = 0.2 \times 509 = 102 \text{ kg/m}^2 \text{ s},$$

and

$$G_l = (1 - x)G = (1 - 0.2) \times 509 = 407 \text{ kg/m}^2 \text{ s}.$$

Therefore

$$\frac{G_g^2}{\rho_g} = \frac{102^2}{10} = 1040 \text{ kg/s}^2 \text{ m},$$

and

$$\frac{G_l^2}{\rho_l} = \frac{407^2}{850} = 195 \text{ kg/s}^2 \text{ m}.$$

Plotting this point on Fig. 2.7, it is found to be well into the annular flow region. Therefore there is little doubt that the flow pattern is *annular*.

This conclusion can be confirmed by evaluating V_g^* (see eqn (11.14).

$$V_g^* = V_g \rho_g^{\frac{1}{2}}/[gd(\rho_l - \rho_g)]^{\frac{1}{2}}, \tag{11.14}$$

and

$$V_g = \text{superficial gas velocity} = \frac{G_g}{\rho_g} = \frac{102}{10} = 10.2 \text{ m/s},$$

then

$$V_g^* = \frac{10.2 \times 10^{\frac{1}{2}}}{[9.81 \times 0.1 \times (850 - 10)]^{\frac{1}{2}}} = 1.12.$$

Annular flow is often said to occur when $V_g^* > 1$, although it can be noted that this criterion for annular flow is only just satisfied.

(b) *Horizontal flow.* The method used is the complex flow pattern of Taitel and Dukler. It is necessary to calculate first the single-phase gas and liquid pressure gradients $(dp/dz)_l$ and $(dp/dz)_g$. This is done by evaluating the single-phase Reynolds number $G_i d/\mu_i$, then the friction factor c_{fi}, and finally the pressure gradient.

If

$$Re_i < 2000, \quad \text{then} \quad c_{fi} = 16/Re_i,$$

if

$$Re_i > 2000, \quad \text{then} \quad c_{fi} = 0.079/Re_i^{\frac{1}{4}},$$

then

$$(dp/dz)_i = -\frac{2c_{fi}G_i^2}{d\rho_i}$$

EXAMPLE 2 BUBBLE FLOW   245

For the gas

$$Re_g = \frac{G_g d}{\mu_g} = \frac{102 \times 0.1}{16 \times 10^{-6}} = 6.37 \times 10^5,$$

$$c_{fg} = \frac{0.079}{(6.37 \times 10^5)^{\frac{1}{4}}} = 0.0028,$$

and

$$(dp/dz)_g = -\frac{2 \times 0.0028 \times 102^2}{0.1 \times 10} = -58.26 \, \text{N/m}^3.$$

For the liquid

$$Re_l = \frac{G_l d}{\mu_l} = \frac{407 \times 0.1}{128 \times 10^{-6}} = 3.18 \times 10^5,$$

$$c_{fl} = \frac{0.079}{(3.18 \times 10^5)^{\frac{1}{4}}} = 0.0033,$$

and

$$(dp/dz)_l = -\frac{2 \times 0.0033 \times 407^2}{0.1 \times 850} = -12.97 \, \text{N/m}^3.$$

The relevant dimensionless groups can now be calculated from eqns (2.5–8)

$$X = \left[\frac{(dp/dz)_l}{(dp/dz)_g}\right]^{\frac{1}{2}} = \left(\frac{-12.97}{-58.26}\right)^{\frac{1}{2}} = 0.47,$$

$$Fr = \frac{G_g}{[\rho_g(\rho_l - \rho_g)dg]^{\frac{1}{2}}} = \frac{102}{[10 \times (850 - 10) \times 0.1 \times 9.81]^{\frac{1}{2}}} = 1.12$$

(note that the definitions of Fr and V_g^* are identical),

$$T = \left[\frac{|(dp/dz)_l|}{g(\rho_l - \rho_g)}\right]^{\frac{1}{2}} = \left[\frac{12.97}{9.81 \times (850 - 10)}\right]^{\frac{1}{2}} = 0.039,$$

and

$$K = Fr\left[\frac{G_l d}{\mu_l}\right]^{\frac{1}{2}} = 1.12 \times \left[\frac{407 \times 0.1}{128 \times 10^{-6}}\right]^{\frac{1}{2}} = 632.$$

The top part of the Fig. 2.8 shows that again the flow pattern is *annular*: the second and third parts of the figure are not needed in this case.

Example 2   Bubble flow

Calculate the variation of the rising velocity of a bubble through stagnant saturated liquid benzene at 1 bar with the size of the bubble.

Solution

The physical properties of saturated benzene at 1 bar are:

  liquid density $\rho_1 = 823 \text{ kg/m}^3$;
  vapour density $\rho_g = 2.74 \text{ kg/m}^3$;
  liquid viscosity $\mu_1 = 321 \times 10^{-6} \text{ N s/m}^2$; and
  surface tension $\sigma = 0.021 \text{ N/m}$.

Using Fig. 3.3, the dimensionless group which can be calculated immediately is the Morton number (eqn (3.8))

$$M = \frac{g\mu_1^4(\rho_1 - \rho_g)}{\rho_1^2\sigma^3} = \frac{9.81 \times (321 \times 10^{-6})^4 \times (823 - 2.74)}{823^2 \times 0.021^3} = 1.36 \times 10^{-11}.$$

From Fig. 3.3 the transition from spherical to wobbling bubbles occurs at $Eo = 0.33$ and $Re = 700$. From eqn (3.6)

$$Eo = \frac{g(\rho_1 - \rho_g)D_e^2}{\sigma}, \tag{3.6}$$

and so when $Eo = 0.33$, $D_e = 9.3 \times 10^{-4} \text{ m} = 0.93 \text{ mm}$. Then, from eqn (3.7),

$$Re = \frac{\rho_1 D_e u_b}{\mu_1}, \tag{3.7}$$

and $Re = 700$, $u_b = 0.29 \text{ m/s}$.

Similarly, the transition to spherical cap bubbles occurs at $Eo = 40$ and $Re = 8000$. These figures lead to the results $D_e = 0.01 \text{ m} = 10 \text{ mm}$, and $u_b = 0.31 \text{ m/s}$.

At very low values of D_e, the rising velocity is given by eqn (3.3)

$$u_b = \frac{D_e^2 g(\rho_1 - \rho_g)}{18\mu_1} = \frac{D_e^2 \times 9.81 \times (823 - 2.74)}{18 \times 321 \times 10^{-6}} = 1.39 \times 10^6 D_e^2,$$

where the units are u_b (m/s) and D_e (m).

In the spherical cap region, the rising velocity is given by eqn (3.15)

$$u_b = 0.71\left[\frac{gD_e(\rho_1 - \rho_g)}{\rho_1}\right]^{\frac{1}{2}} = 0.71 \times \left[\frac{9.81 \times D_e \times (823 - 2.74)}{823}\right]^{\frac{1}{2}}$$

$$= 2.22 D_e^{\frac{1}{2}}.$$

The values of u_b and the variation with D_e are shown on Fig. 25.1. An equation for the bubble rise velocity is also given in Chapter 7. This is

$$u_b = 1.4\left[\frac{\sigma g(\rho_1 - \rho_g)}{\rho_1^2}\right]^{\frac{1}{4}}. \tag{7.36}$$

EXAMPLE 3   PLUG FLOW   247

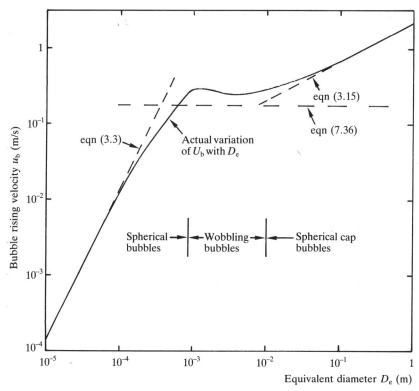

Fig. 25.1. Values of U_b and its variation with D_e for Example 2.

Substituting the values in this equation gives

$$u_b = 0.176 \, \text{m/s}.$$

As can be seen from Fig. 25.1, this value represents, approximately, the rise velocity in the region where the bubble shape is changing from spherical to wobbling to spherical cap.

Example 3   Plug flow

Calculate the plug rising velocity for an air plug rising through a viscous liquid (see properties below) in a vertical tube of diameter 0.01 m. The physical properties are:

liquid density $\rho_l = 100 \, \text{kg/m}^3$;
gas density $\rho_g = 1.3 \, \text{kg/m}^3$;
liquid viscosity $\mu_l = 0.1 \, \text{N s/m}^2$;   and
surface tension $\sigma = 0.06 \, \text{N/m}$.

Solution

Calculate the Morton number from eqn (3.8) and the Eötvös number (using the tube diameter in place of the bubble diameter) from eqn (3.6).

$$M = \frac{g\mu_1^4(\rho_1 - \rho_g)}{\rho_1^2 \sigma^3} = \frac{9.81 \times 0.1^4 \times (1000 - 1.3)}{1000^2 \times 0.06^3}$$

$$= 4.5 \times 10^{-3}$$

$$Eo = \frac{g(\rho_1 - \rho_g)d^2}{\sigma} = \frac{9.81 \times (1000 - 1.3) \times 0.01^2}{0.06}$$

$$= 16.3.$$

Then, from Fig. 3.8, $Fr = 0.09$. Note that here both the viscosity and surface tension effects cannot be neglected and it would be entirely wrong to assume that $Fr = 0.35$. Then, from eqn (3.16), the definition of the Froude number is

$$u_p = Fr(gd)^{\frac{1}{2}}\left(\frac{\rho_1 - \rho_g}{\rho_1}\right)^{\frac{1}{2}}$$

$$= 0.09 \times (9.81 \times 0.01)^{\frac{1}{2}} \times \left(\frac{1000 - 1.3}{1000}\right)^{\frac{1}{2}} = 0.028 \text{ m/s}.$$

Example 4   Two-phase pressure drop and void fraction

A steam–water mixture of quality 0.1 at 5 bar flows up a smooth vertical round tube of diameter 0.05 m. The total flow rate is 0.6 kg/s. Calculate:

(*a*) the homogeneous void fraction and the corresponding gravitational pressure gradient;
(*b*) the homogeneous frictional pressure gradient;
(*c*) the void fraction and gravitational pressure gradient using the separated flow model; and
(*d*) the frictional pressure gradient using the separated flow model.

The required physical properties are:

liquid density $\rho_1 = 915 \text{ kg/m}^3$;
gas density $\rho_g = 2.67 \text{ kg/m}^3$;
liquid viscosity $\mu_1 = 180 \times 10^{-6} \text{ N s/m}^2$;
gas viscosity $\mu_g = 14 \times 10^{-6} \text{ N s/m}^2$;   and
surface tension $\sigma = 0.0487 \text{ N/m}$.

EXAMPLE 4 TWO-PHASE PRESSURE DROP   249

Solution

(a) *Homogeneous void fraction and gravitational pressure gradient.* From eqn (5.5)

$$\alpha_h = \frac{1}{1 + \left(\dfrac{1-x}{x}\dfrac{\rho_g}{\rho_l}\right)} = \frac{1}{1 + \left(\dfrac{1-0.1}{0.1} \times \dfrac{2.67}{915}\right)} = 0.9744.$$

The homogeneous density can be found from eqn (5.11)

$$\rho_h = \rho_g \alpha_h + \rho_l(1-\alpha_h) = 2.67 \times 0.9744 + 915 \times (1 - 0.9744)$$
$$= 26.0 \text{ kg/m}^3.$$

Then use eqn (5.8). For vertical upflow $\theta = 90°$

$$-\left(\frac{dp}{dz}\right)_g = \rho_h g \sin \theta = 26.0 \times 9.81 \times 1$$
$$= 255 \text{ N/m}^3.$$

(b) *Homogeneous frictional pressure gradient.* First, calculate the homogeneous viscosity from eqn (5.13)

$$\frac{1}{\mu_h} = \frac{x}{\mu_g} + \frac{1-x}{\mu_l} = \frac{0.1}{14 \times 10^{-6}} + \frac{1-0.1}{180 \times 10^{-6}}$$

so

$$\mu_h = 82.3 \times 10^{-6} \text{ N s/m}^2.$$

The total mass flux G is

$$G = \frac{\text{mass flow rate}}{\text{cross-sectional area}} = \frac{0.6}{\pi \times 0.05^2/4} = 305.6 \text{ kg/m}^2 \text{ s}.$$

The mixture Reynolds number (see eqn (5.10)) is

$$Re_h = \frac{Gd}{\mu_h} = \frac{305.6 \times 0.05}{82.3 \times 10^{-6}} = 1.86 \times 10^5.$$

Using the friction factor equations from Example 1,

$$c_f = \frac{0.079}{(1.86 \times 10^5)^{\frac{1}{4}}} = 0.003\,81.$$

Then, also as in Example 1,

$$-\left(\frac{dp}{dz}\right)_f = \frac{2c_f G^2}{d\rho_h} = \frac{2 \times 0.003\,81 \times 305.6^2}{0.05 \times 26.0}$$
$$= 547 \text{ N/m}^3.$$

(c) *Separated flow void fraction and gravitational pressure gradient.* The recommended method for the void fraction is detailed in Appendix A. Only the intermediate answers, not the workings, are given here.

Equation (A7): $Re = 8.49 \times 10^5$.

Equation (A8): $We = 105$.

Equation (A5): $E_1 = 0.426$.

Equation (A6): $E_2 = 1.7 \times 10^{-3}$.

Equation (A4): $\beta = 0.9744$ (note β is necessarily equal to α_h).

Equation (A3): $y = 38.1$.

Equation (A2): $S = 3.54$. If, in evaluating this equation, it becomes necessary to find the square root of a negative number, then put $S = 1$.

Equation (A1): $\alpha = 0.9148$.

Then, as before,

$$\rho = \rho_g \alpha + \rho_l(1 - \alpha) = 2.67 \times 0.9148 + 915 \times (1 - 0.9148)$$
$$= 80.4 \text{ kg/m}^3,$$

and

$$-\left(\frac{\mathrm{d}p}{\mathrm{d}z}\right)_g = \rho g \sin \theta = 80.4 \times 9.81 \times 1 = 789 \text{ N/m}^3.$$

(d) *Separated flow frictional pressure gradient.* The recommended method is detailed in Appendix B. Only the intermediate answers, not the workings, are given here.

From previous calculations, or from eqn (B8), $\rho_h = 26.0 \text{ kg/m}^3$.

Equation (B7): $We = 3757$;

Equation (B6); $Fr = 281.7$;

Equation (B5): $H = 118$; and

Equation (B4): $F = 0.162$.

In order to calculate $c_{f_{go}}$ and $c_{f_{lo}}$, first calculate the Reynolds numbers as if the total flow were gas and liquid.

$$Re_{go} = \frac{Gd}{\mu_g} = \frac{305.6 \times 0.05}{14 \times 10^{-6}} = 1.09 \times 10^6$$

$$c_{fgo} = \frac{0.079}{Re_{go}^{\frac{1}{4}}} = \frac{0.079}{(1.09 \times 10^6)^{\frac{1}{4}}} = 0.002\,44$$

$$Re_{lo} = \frac{Gd}{\mu_l} = \frac{305.6 \times 0.05}{180 \times 10^{-6}} = 8.49 \times 10^4$$

$$c_{flo} = \frac{0.079}{Re_{lo}^{\frac{1}{4}}} = \frac{0.079}{(8.49 \times 10^4)^{\frac{1}{4}}} = 0.004\,63$$

EXAMPLE 5 CRITICAL TWO-PHASE FLOW 251

If the Reynolds numbers had been less than 2000, then $c_f = 16/Re$ would have been used.

Equation (B3): $E = 2.62$

Equation (B2): $\phi_{lo}^2 = 38.64$

The single-phase liquid pressure gradient using the total mass flow $(dp/dz)_{lo}$ can be calculated from

$$-\left(\frac{dp}{dz}\right)_{lo} = \frac{2c_{flo}G^2}{d\rho_1} = \frac{2 \times 0.004\,63 \times 305.6^2}{0.05 \times 915}$$

$$= 18.9\ \text{N/m}^3.$$

From eqn (B1)

$$-\left(\frac{dp}{dz}\right)_f = -\left(\frac{dp}{dz}\right)_{lo} \phi_{lo}^2 = 18.9 \times 38.64$$

$$= 730\ \text{T/m}^3.$$

The results are summarized in Table 25.1.

TABLE 25.1. *Summary of Example 4 results*

Model	Void fraction	$-\left(\dfrac{dp}{dz}\right)_g$	$-\left(\dfrac{dp}{dz}\right)_f$
Homogeneous	0.9744	255 N/m³	547 N/m³
Separated	0.9148	789 N/m³	730 N/m³

This example is at low pressure and low mass flux, so the separated model produces the more accurate results.

Example 5   Critical two-phase flow

Find the critical flow rate through a nozzle of diameter 0.01 m when the upstream conditions are a pressure of 10 bar and a quality of 0.02:

(a) for benzene using the homogeneous frozen-flow model;

(b) for water using the homogeneous frozen-flow model, and

(c) for water using the homogeneous equilibrium model.

The physical property data at 10 bar are given in Table 25.2.

TABLE 25.2. *Physical property data for Example 5*

Property		Benzene	Water
Liquid density	ρ_1	695 kg/m³	887 kg/m³
Vapour density	ρ_g	24 kg/m³	5.15 kg/m³
Ratio of specific heats	γ	1.15	1.3

Solution

(a) For homogeneous frozen flow, use eqns (8.22) and (8.20), remembering that specific volume is the reciprocal of density, and that the correct units for pressure are N/m^2 (1 bar $= 10^5 N/m^2$)

$$v = (1 - x)v_{l0} + xv_{g0}\left(\frac{2}{\gamma + 1}\right)^{\frac{1}{1-\gamma}}$$

$$= (1 - 0.02)\frac{1}{695} + 0.02 \times \frac{1}{24} \times \left(\frac{2}{1.15 + 1}\right)^{\frac{1}{1-1.15}}$$

$$= 2.76 \times 10^{-3} \, m^3/kg,$$

and

$$G_c = \left[\frac{xv_{g0}p_0\gamma}{v^2}\frac{2}{(\gamma + 1)}\right]^{\frac{1}{2}}$$

$$= \left[\frac{0.02 \times \dfrac{1}{24} \times 10^6 \times 1.15 \times 2}{(2.76 \times 10^{-3})^2 \times (1.15 + 1)}\right]^{\frac{1}{2}} = 10\,800 \, kg/m^2 \, s.$$

The cross-sectional area of the nozzle is $\pi(0.01)^2/4 = 7.85 \times 10^{-5} \, m^2$, so the critical flow rate is $10\,800 \times 7.85 \times 10^{-5} = 0.85$ kg/s.

(b) Repeating the calculation with water properties

$$v = (1 - 0.02)\frac{1}{887} + 0.02 \times \frac{1}{5.15} \times \left(\frac{2}{1.3 + 1}\right)^{\frac{1}{1-1.3}}$$

$$= 7.29 \times 10^{-3} \, m^3/kg,$$

and

$$G_c = \left[\frac{0.02 \times \dfrac{1}{5.15} \times 10^6 \times 1.3 \times 2}{(7.29 \times 10^{-3})^2 \times (1.3 + 1)}\right]^{\frac{1}{2}} = 9090 \, kg/m^2 \, s.$$

Therefore critical flow rate $= 9090 \times 7.85 \times 10^{-5} = 0.71$ kg/s.

(c) For the homogeneous equilibrium model a relatively large amount of thermodynamic information is required (see § 8.4). However, for the special case of water Fig. 8.3 can be used.

h_0 is the stagnation enthalpy (see eqn 8.25). At 10 bar the liquid and vapour saturation enthalpies are:

$h_l = 762.6 \times 10^3 \, J/kg;$ and
$h_g = 2776.2 \times 10^3 \, J/kg.$

EXAMPLE 6 FLOODING 253

Therefore

$$h_0 = (1 - x)h_1 + xh_g$$
$$= (1 - 0.02) \times 762.6 \times 10^3 + 0.02 \times 2776.2 \times 10^3$$
$$= 802.9 \times 10^3 \text{ J/kg}.$$

Using this value for h_0, $p_0 = 10$ bar and Fig. 8.3, then $G_c = 5700 \text{ kg/m}^2$ s. This value obtained from the figure is only approximate: a more accurate value could be obtained by detailed calculations, but the effort involved is considerable. The critical mass flow rate is therefore

$$5700 \times 7.85 \times 10^{-5} = 0.45 \text{ kg/s}.$$

As stated in § 8.5, the homogeneous frozen model is not only easier to use, but also gives better predictions of the critical flow rate.

Example 6 Flooding

0.03 kg/s of saturated water at 1 bar flows down the walls of a tube of diameter 0.03 m. Find the upwards mass flow rate of saturated steam at the same pressure which is necessary to cause flooding:

(a) using a Wallis equation with $C = 0.8$;
(b) using a gas Kutateladze number of 3.2; and
(c) using eqn (11.18).

The physical properties are:

liquid density $\rho_1 = 958 \text{ kg/m}^3$;
gas density $\rho_g = 0.59 \text{ kg/m}^3$;
liquid viscosity $\mu_1 = 283 \times 10^{-6} \text{ N s/m}^2$;   and
surface tension $\sigma = 0.0588 \text{ N/m}$.

Solution

(a) The tube cross-sectional area $= \pi(0.03)^2/4$

$$= 7.07 \times 10^{-4} \text{ m}^2.$$

The liquid superficial velocity

$$V_1 = \frac{\text{mass flow rate}}{\text{area} \times \text{density}} = \frac{0.03}{(7.07 \times 10^{-4}) \times 958} = 0.044 \text{ m/s}.$$

Then, using eqn (11.13),

$$V_1^* = \frac{V_1 \rho_1^{\frac{1}{2}}}{[gd(\rho_1 - \rho_g)]^{\frac{1}{2}}} = \frac{0.044 \times 958^{\frac{1}{2}}}{[9.81 \times 0.03 \times (958 - 0.59)]^{\frac{1}{2}}}$$
$$= 0.081.$$

The Wallis correlation, eqn (11.12), is

$$V_1^{*\frac{1}{2}} + V_g^{*\frac{1}{2}} = 0.8,$$

therefore $V_g^* = 0.265$ and, from eqn (11.14),

$$V_g = \frac{V_g^*[gd(\rho_1 - \rho_g)]^{\frac{1}{2}}}{\rho_g^{\frac{1}{2}}}$$

$$= \frac{0.265 \times [9.81 \times 0.03 \times (958 - 0.59)]^{\frac{1}{2}}}{0.59^{\frac{1}{2}}}$$

$$= 5.8 \text{ m/s}.$$

The mass flow rate is velocity × area × gas density

$$= 5.8 \times (7.07 \times 10^{-4}) \times 0.59 = 2.4 \times 10^{-3} \text{ kg/s}.$$

(b) The gas Kutateladze number is given by eqn (11.15); rearranging this equation

$$V_g = \frac{K_g[g\sigma(\rho_1 - \rho_g)]^{\frac{1}{4}}}{\rho_g^{\frac{1}{2}}}$$

$$V_g = \frac{3.2 \times [9.81 \times 0.0588 \times (958 - 0.59)]^{\frac{1}{4}}}{0.59^{\frac{1}{2}}}$$

$$= 20.2 \text{ m/s};$$

and the corresponding mass flow rate is 8.4×10^{-3} kg/s.

(c) Using eqns (11.20), (11.19) and (11.18)

$$Fr = V_1 \frac{\pi}{4} \left[\frac{g(\rho_1 - \rho_g)}{\sigma} \right]^{\frac{1}{4}} = 0.044 \times \frac{\pi}{4} \times \left[\frac{9.81 \times (958 - 0.59)}{0.0588} \right]^{\frac{1}{4}}$$

$$= 0.691;$$

$$Bo = \frac{d^2 g(\rho_1 - \rho_g)}{\sigma} = \frac{0.03^2 \times 9.81 \times (958 - 0.59)}{0.0588}$$

$$= 143.8;$$

$$K_g = 0.286 Bo^{0.26} Fr^{-0.22} \left[1 + \frac{\mu_1}{\mu_w} \right]^{-0.18}$$

$$= 0.286 \times 143.8^{0.26} \times 0.691^{-0.22} \times \left[1 + \frac{283 \times 10^{-6}}{10^{-3}} \right]^{-0.18}$$

$$= 1.08.$$

Then, the corresponding values of V_g and the gas flow rate are 6.82 m/s and 2.8×10^{-3} kg/s.

The answers are summarized in Table 25.3.

EXAMPLE 7   POOL BOILING CURVE   255

TABLE 25.3.   *Summary of results for Example* 6

Correlation	Mass flow rate (kg/s)
Wallis correlation ($C = 0.8$)	2.4×10^{-3}
Kutateladze number ($K_g = 3.2$)	8.4×10^{-3}
Optimized correlation	2.8×10^{-3}

It is probable in this case that the constant K_g correlation gives very poor results.

Example 7   Pool boiling curve

Draw the pool boiling curve for benzene at 1 bar. The saturation physical properties are:

molecular weight	$M = 78$;
liquid density	$\rho_l = 823 \text{ kg/m}^3$;
vapour density	$\rho_g = 2.74 \text{ kg/m}^3$;
latent heat of vaporization	$\lambda = 398 \times 10^3 \text{ J/kg}$;
liquid specific heat	$C_{pl} = 1880 \text{ J/kg K}$;
vapour specific heat	$C_{pg} = 1290 \text{ J/kg K}$;
liquid viscosity	$\mu_l = 321 \times 10^{-6} \text{ N s/m}^2$;
vapour viscosity	$\mu_g = 9.3 \times 10^{-6} \text{ N s/m}^2$;
liquid thermal conductivity	$k_l = 0.131 \text{ W/m K}$;
vapour thermal conductivity	$k_g = 0.015 \text{ W/m K}$;
surface tension	$\sigma = 0.021 \text{ N/m}$;
saturation temperature	$T_{sat} = 353 \text{ K } (80\,°C)$; and
critical pressure	$p_c = 49 \text{ bar}$.

The boiling takes place on a flat, horizontal surface with a cavity diameter of 10 μm (so cavity radius $R = 5 \times 10^{-6}$ m), and a surface roughness ε of 1 μm.

Solution

The solution can be divided into a number of distinct parts (see Fig. 25.2).

(a) onset of nucleate boiling (see A in Fig. 25.2),
(b) nucleate boiling region, A to B;
(c) critical heat flux point, B;
(d) transition boiling, B to C;
(e) minimum film boiling point, C; and
(f) film boiling region, C onwards.

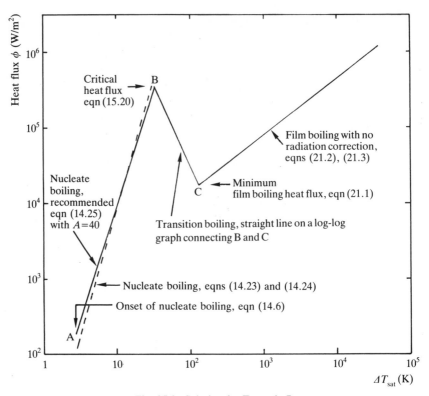

Fig. 25.2. Solution for Example 7.

(a) *Onset of nucleate boiling* (from eqn (14.6)).

$$\Delta T_{sat} = \frac{2\sigma T_{sat}}{\rho_g \lambda R} = \frac{2 \times 0.021 \times 353}{2.74 \times 398 \times 10^3 \times 5 \times 10^{-6}} = 2.7 \text{ K}.$$

Thus nucleate boiling will start when the surface temperature reaches $T_{sat} + \Delta T_{sat} = 80 + 2.7 = 82.7\,°C$. One co-ordinate ($\Delta T_{sat}$) of the point A on Fig. 25.2 is now known.

(b) *Nucleate boiling region.* The reduced pressure p_R is

$$p_R = \frac{\text{system pressure}}{\text{critical pressure}} = \frac{1}{49} = 0.0204,$$

and so, from eqn (14.25), taking a value of 40 for the constant A

$$h = 40 p_R^{(0.12 - \log_{10}\varepsilon)}(-\log_{10} p_R)^{-0.55} M^{-0.5} \phi^{\frac{2}{3}}$$

$$h = 40 \times 0.0204^{(0.12 - \log_{10}1)} \times (-\log_{10} 0.0204)^{-0.55} \times 78^{-0.5} \times \phi^{\frac{2}{3}}$$

$$h = 2.13 \phi^{\frac{2}{3}},$$

EXAMPLE 7 POOL BOILING CURVE 257

where h is the boiling heat transfer coefficient (W/m^2 K) and ϕ is the heat flux (W/m^2). Eliminating h by using $h = \phi / \Delta T_{sat}$

$$\phi = 2.13^3 \Delta T_{sat}^3 = 9.62 \Delta T_{sat}^3.$$

This is the equation of the nucleate boiling line and is shown on Fig. 25.2. The point A can now be located by substituting $\Delta T_{sat} = 2.7$ K

$$\phi = 9.62 \times 2.7^3 = 189 \text{ W/m}^2.$$

If the Mostinski method is used (eqns (14.23) and (14.24)), then the result is slightly different

$$h = 1.50 \phi^{0.7}, \quad \text{or} \quad \phi = 3.86 \Delta T_{sat}^{3.33}.$$

(c) *Critical heat flux.* If the heater is of a reasonable size, then eqn (15.20) can be used:

$$\phi = 0.149 \lambda \rho_g^{\frac{1}{2}} [\sigma(\rho_1 - \rho_g)g]^{\frac{1}{4}}$$
$$= 0.149 \times 398 \times 10^3 \times 2.74^{\frac{1}{2}} \times [0.021 \times (823 - 2.74) \times 9.81]^{\frac{1}{4}}$$
$$= 3.54 \times 10^5 \text{ W/m}^2.$$

This critical heat flux point is marked B on Fig. 25.2. It is also on the nucleate boiling line.

(d) The transition boiling line cannot be located until the minimum film boiling point has been found.

(e) *Minimum film boiling point* (from eqn (21.1))

$$\phi_{min} = 0.13 \lambda \rho_g \left[\frac{\sigma g(\rho_1 - \rho_g)}{(\rho_1 + \rho_g)^2} \right]^{\frac{1}{4}}$$
$$= 0.13 \times 398 \times 10^3 \times 2.74 \times \left[\frac{0.021 \times 9.81 \times (823 - 2.74)}{(823 + 2.74)^2} \right]^{\frac{1}{4}}$$
$$= 1.78 \times 10^4 \text{ W/m}^2.$$

Note that this is only 5 % of the critical heat flux. The minimum film boiling point cannot yet be located on the boiling curve as only one co-ordinate (ϕ) is yet known. The other co-ordinate (ΔT_{sat}) is calculated from the fact that the point is on the film boiling line.

(f) *Film boiling region* (from eqns (21.3) and (21.2))

$$l = 2\pi \left[\frac{\sigma}{g(\rho_1 - \rho_g)} \right]^{\frac{1}{2}} = 2\pi \times \left[\frac{0.021}{9.81 \times (823 - 2.74)} \right]^{\frac{1}{2}} = 0.010 \text{ m};$$

$$h = 0.62 \left[\frac{\rho_g(\rho_1 - \rho_g)g\lambda k_g^3}{\mu_g \Delta T_{sat} l} \right]^{\frac{1}{4}}$$
$$= 0.62 \times \left[\frac{2.74 \times (823 - 2.74) \times 9.81 \times 398 \times 10^3 \times 0.015^3}{9.3 \times 10^{-6} \times \Delta T_{sat} \times 0.01} \right]^{\frac{1}{4}}$$
$$= \frac{466}{\Delta T_{sat}^{\frac{1}{4}}}.$$

Then, eliminating h by using $h = \phi/\Delta T_{\text{sat}}$

$$\phi = 466\Delta T_{\text{sat}}^{\frac{3}{4}}.$$

Substituting the heat flux at the minimum film boiling point ($\phi_{\text{min}} = 1.78 \times 10^4 \, \text{W/m}^2$), the value of ΔT_{sat} at that point can be found

$$\Delta T_{\text{sat}} = \left[\frac{1.78 \times 10^4}{466} \right]^{\frac{4}{3}} = 128.6 \, \text{K}.$$

The complete boiling curve can now be drawn. The transition boiling line is assumed to be the straight line (on a graph of $\log \phi$ against $\log \Delta T_{\text{sat}}$) connecting the critical heat flux point and the minimum film boiling point (see Fig. 25.2).

Note that no radiation correction was made for the film boiling region. At high temperatures, the radiation heat transfer is very important. Example 9(b) shows the general method of the correction.

Example 8 Flow boiling

Water at 20 bar flows up a vertical round tube of internal diameter 0.02 m. The tube is 6 m long, and the water enters at 120 °C. The heat flux is 250 kW/m² and the water mass flux is 500 kg/m² s. Calculate:

(a) the wall temperature difference ΔT_{sat} required for the onset of nucleate boiling;
(b) the boiling heat transfer coefficient at the top of the tube;
(c) the critical heat flux using the Bowring correlation; and
(d) the critical heat flux using the Katto correlation.

The physical properties are:

liquid density	$\rho_1 = 850 \, \text{kg/m}^3$;
vapour density	$\rho_g = 10.0 \, \text{kg/m}^3$;
saturated liquid enthalpy at 20 bar	$= 908.6 \times 10^3 \, \text{J/kg}$;
liquid enthalpy at 120 °C	$= 504.0 \times 10^3 \, \text{J/kg}$;
latent heat of vaporisation at 20 bar	$\lambda = 1890 \times 10^3 \, \text{J/kg}$;
liquid viscosity	$\mu_1 = 127 \times 10^{-6} \, \text{N s/m}^2$;
vapour viscosity	$\mu_g = 16 \times 10^{-6} \, \text{N s/m}^2$;
liquid thermal conductivity	$k_1 = 0.65 \, \text{W/m K}$;
vapour thermal conductivity	$k_g = 0.040 \, \text{W/m K}$;
liquid specific heat	$C_{\text{pl}} = 4560 \, \text{J/kg K}$;
vapour specific heat	$C_{\text{pg}} = 3060 \, \text{J/kg K}$;
surface tension	$\sigma = 0.035 \, \text{N/m}$; and
saturation temperature T_{sat} at 20 bar	$= 485 \, \text{K} \, (= 212 \, ^\circ\text{C})$.

EXAMPLE 8 FLOW BOILING 259

Solution

(a) The onset of nucleate boiling in a flow situation is given by eqn (16.10)

$$\Delta T_{sat} > \left(\frac{8\sigma T_{sat}\phi}{\lambda \rho_g k_1}\right)^{\frac{1}{2}}.$$

Therefore

$$\Delta T_{sat} > \left[\frac{8 \times 0.035 \times 485 \times 250 \times 10^3}{1890 \times 10^3 \times 10 \times 0.65}\right]^{\frac{1}{2}},$$

or

$$\Delta T_{sat} > 1.7 \text{ K}.$$

Thus nucleate boiling will begin when the wall temperature is 1.7 K above the saturation temperature.

(b) *Heat transfer coefficient.* First, it is necessary to find the quality at the required position (the end of the tube) by means of a heat balance.

Heat supplied to the tube $= \pi d L \phi$.

Heat taken up by the water = mass flow rate × change in specific

$$\text{enthalpy} = \frac{\pi d^2}{4} G(x\lambda + \Delta h_s).$$

Δh_s is the inlet subcooling, that is the amount by which the enthalpy of the liquid is below the saturation enthalpy. Here, $\Delta h_s = 908.6 \times 10^3 - 504.0 \times 10^3 = 404.6 \times 10^3$ J/kg. Equating the heat supplied to the heat taken up

$$\pi d L \phi = \frac{\pi d^2}{4} G(x\lambda + \Delta h_s),$$

or

$$x = \frac{4L\phi}{dG\lambda} - \frac{\Delta h_s}{\lambda}$$

$$= \frac{4 \times 6 \times 250 \times 10^3}{0.02 \times 500 \times 1890 \times 10^3} - \frac{404.6 \times 10^3}{1890 \times 10^3} = 0.103.$$

Calculate Re_1 from eqn (16.18)

$$Re_1 = \frac{G(1-x)d}{\mu_1} = \frac{500 \times (1-0.103) \times 0.02}{127 \times 10^{-6}} = 70\,600.$$

Calculate Pr_1 from eqn (16.19)

$$Pr_1 = \frac{\mu_1 C_{pl}}{k_1} = \frac{127 \times 10^{-6} \times 4560}{0.65} = 0.891.$$

Calculate Nu_1 from eqn (16.16)

$$Nu_1 = 0.023Re_1^{0.8}Pr_1^{0.4} = 0.023 \times 70\,600^{0.8} \times 0.891^{0.4} = 166.$$

Calculate h_1 from eqn (16.17)

$$h_1 = \frac{Nu_1 k_1}{d} = \frac{166 \times 0.65}{0.02} = 5400 \text{ W/m}^2 \text{ K.}$$

Calculate X from eqn (16.21)

$$X = \left(\frac{1-x}{x}\right)^{0.9} \left(\frac{\rho_g}{\rho_l}\right)^{0.5} \left(\frac{\mu_l}{\mu_g}\right)^{0.1}$$

$$= \left(\frac{1-0.103}{0.103}\right)^{0.9} \times \left(\frac{10}{850}\right)^{0.5} \times \left(\frac{127 \times 10^{-6}}{16 \times 10^{-6}}\right)^{0.1} = 0.94.$$

From Fig. 16.8, when $X = 0.94$, $F = 2.8$.
Calculate h_{FC} from eqn (16.15)

$$h_{FC} = h_1 F = 5400 \times 2.8 = 15\,100 \text{ W/m}^2 \text{ K.}$$

Calculate Re_{TP} from eqn (16.23)

$$Re_{TP} = Re_1 F^{1.25} = 70\,600 \times 2.8^{1.25} = 2.56 \times 10^5.$$

From Fig. 16.9, when $Re_{TP} = 2.56 \times 10^5$, $S = 0.16$.
Calculate h_{FZ} from eqn (16.13) in terms of ΔT_{sat} and Δp_{sat}

$$h_{FZ} = \frac{0.001\,22\Delta T_{sat}^{0.24}\Delta p_{sat}^{0.75}C_{pl}^{0.45}\rho_l^{0.49}k_1^{0.79}}{\sigma^{0.5}\lambda^{0.24}\mu_l^{0.29}\rho_g^{0.24}}$$

$$= \frac{0.001\,22\Delta T_{sat}^{0.24}\Delta p_{sat}^{0.75}4560^{0.45} \times 850^{0.49} \times 0.65^{0.79}}{0.035^{0.5} \times (1890 \times 10^3)^{0.24} \times (127 \times 10^{-6})^{0.29} \times 10^{0.24}}$$

$$= 1.36\Delta T_{sat}^{0.24}\Delta p_{sat}^{0.75}.$$

Now the procedure is iterative: ΔT_{sat} is guessed, Δp_{sat} is found, then h_{FZ}, then h_{NB} from eqn (16.12)

$$h_{NB} = Sh_{FZ};$$

and then the overall boiling coefficient is found from eqn (16.11)

$$h_B = h_{NB} + h_{FC}.$$

Finally, a new value of T_{sat} is calculated from

$$\Delta T_{sat} = \frac{\phi}{h_B}.$$

Going through this procedure, starting with an initial guess of $\Delta T_{sat} = 5$ K: Δp_{sat} is the vapour pressure at $(T_{sat} + 5)$ minus vapour pressure at

EXAMPLE 8 FLOW BOILING 261

T_{sat}. Note that the units of Δp_{sat} must be N/m^2. Here,

$$\Delta p_{sat} = 2.06 \text{ bar} = 2.06 \times 10^5 \text{ N/m}^2.$$

Therefore

$$h_{FZ} = 1.36 \Delta T_{sat}^{0.24} \Delta p_{sat}^{0.75}$$
$$= 1.36 \times 5^{0.24} \times (2.06 \times 10^5)^{0.75} = 19\,350 \text{ W/m}^2 \text{ K},$$

and

$$h_{NB} = Sh_{FZ} = 0.16 \times 19\,350 = 4000 \text{ W/m}^2 \text{ K},$$
$$h_B = h_{NB} + h_{FC} = 4000 + 15\,100 = 19\,100 \text{ W/m}^2 \text{ K},$$

and

$$\Delta T_{sat} = \frac{\phi}{h_B} = \frac{250 \times 10^3}{19\,100} = 13.1 \text{ K}.$$

The results of repeating the iteration are given in Table 25.4. Thus the final result is that the boiling heat transfer coefficient is $22\,200 \text{ W/m}^2 \text{ K}$, or $22.2 \text{ kW/m}^2 \text{ K}$.

TABLE 25.4 *Results of the iteration in Example 8(b)*

Quantity	2nd iteration	3rd iteration	4th iteration	5th iteration
ΔT_{sat} (K)	13.1	9.9	11.5	11.2
Δp_{sat} (N/m^2)	5.73×10^5	4.47×10^5	4.97×10^5	4.83×10^5
h_{FZ} (W/m^2 K)	52 500	40 800	45 700	44 500
h_{NB} (W/m^2 K)	8400	6500	7300	7100
h_B (W/m^2 K)	25 300	21 600	22400	22 200
ΔT_{sat} (K)	9.9	11.5	11.2	11.3

(c) *Critical heat flux—method for water.* This method is given in detail in Appendix C. Note that the conditions for the applicability of the method are all satisfied, except that the actual tube length (6 m) is not within the stated range of 0.15 m to 3.7 m. This will inevitably reduce the accuracy of the method.

Calculate p' from eqn (C4)

$$p' = \frac{p}{69} = \frac{20}{69} = 0.290 \quad (p \text{ must be in bar for this equation}).$$

Calculate n from eqn (C5)

$$n = 2.0 - 0.5p' = 2.0 - (0.5 \times 0.290) = 1.855.$$

Then, as $p' < 1$, work out F_1, F_2, F_3, F_4 from eqns (C6), (C7), (C8), and

(C9). The detailed working is not given here: the results are

$F_1 = 0.478,$
$F_2 = 0.440,$
$F_3 = 0.400,$ and
$F_4 = 0.052.$

A' is then calculated from eqn (C2)

$$A' = \frac{0.5792\lambda dGF_1}{1 + 0.0143F_2 d^{0.5}G} = \frac{0.5792 \times (1890 \times 10^3) \times 0.02 \times 500 \times 0.478}{1 + (0.0143 \times 0.440 \times 0.02^{0.5} \times 500)}$$

$$= 3.62 \times 10^6.$$

C' is calculated from eqn (C3)

$$C' = \frac{0.077F_3 dG}{1 + 0.347F_4(G/1356)^n} = \frac{0.077 \times 0.400 \times 0.02 \times 500}{1 + [0.347 \times 0.052 \times (500/1356)^{1.855}]}$$

$$= 0.307.$$

The critical heat flux is then calculated from eqn (C1)

$$\phi_c = \frac{A' + 0.25dG\Delta h_s}{C' + L} = \frac{(3.62 \times 10^6) + [0.25 \times 0.02 \times 500 \times (404.6 \times 10^3)]}{0.307 + 6}$$

$$= 7.34 \times 10^5 \text{ W/m}^2.$$

The quality at the end of the tube is then given by

$$x = \frac{4L\phi_c}{dG\lambda} - \frac{\Delta h_s}{\lambda} = \frac{4 \times 6 \times (7.34 \times 10^5)}{0.02 \times 500 \times (1890 \times 10^3)} - \frac{(404.6 \times 10^3)}{(1890 \times 10^3)}$$

$$= 0.72.$$

This should of course always be less than unity.

(d) *Critical heat flux—method for a general fluid.* This method is given in detail in Appendix D.

First, evaluate the dimensionless groups L', R', and W' from eqns (D2), (D3), and (D4)

$$L' = \frac{L}{d} = \frac{6}{0.02} = 300,$$

$$R' = \frac{\rho_g}{\rho_1} = \frac{10}{850} = 0.011\,8,$$

and

$$W' = \frac{\sigma\rho_1}{G^2 L} = \frac{0.035 \times 850}{500^2 \times 6} = 1.98 \times 10^{-5}.$$

Then calculate C from eqns (D10–12); here (D12) is the relevant

EXAMPLE 8   FLOW BOILING   263

equation

$$C = 0.34.$$

$X_1, X_2, X_3, X_4,$ and X_5 are calculated from eqns (D5–9)

$$X_1 = \frac{CW'^{0.043}}{L'} = \frac{0.34 \times (1.98 \times 10^{-5})^{0.043}}{300} = 7.11 \times 10^{-4},$$

$$X_2 = \frac{0.1R'^{0.133}W'^{0.333}}{1 + 0.0031L'} = \frac{0.1 \times 0.0118^{0.133} \times (1.98 \times 10^{-5})^{0.333}}{1 + (0.0031 \times 300)}$$

$$= 7.79 \times 10^{-4},$$

$$X_3 = \frac{0.098R'^{0.133}W'^{0.433}L'^{0.27}}{1 + 0.0031L'}$$

$$= \frac{0.098 \times 0.0118^{0.133} \times (1.98 \times 10^{-5})^{0.433}300^{0.27}}{1 + (0.0031 \times 300)}$$

$$= 12.06 \times 10^{-4},$$

$$X_4 = \frac{0.0384R'^{0.6}W'^{0.173}}{1 + 0.28W'^{0.233}L'} = \frac{0.0384 \times 0.0118^{0.6} \times (1.98 \times 10^{-5})^{0.173}}{1 + (0.28 \times (1.98 \times 10^{-5})^{0.233} \times 300)}$$

$$= 0.53 \times 10^{-4},$$

and

$$X_5 = \frac{0.234R'^{0.513}W'^{0.433}L'^{0.27}}{1 + 0.0031L'}$$

$$= \frac{0.234 \times 0.0118^{0.513} \times (1.98 \times 10^{-5})^{0.433}300^{0.27}}{1 + (0.0031 \times 300)}$$

$$= 5.33 \times 10^{-4}.$$

$K_1, K_2,$ and K_3 are calculated from eqns (D13–15)

$$K_1 = \frac{0.261}{CW'^{0.043}} = \frac{0.261}{0.34 \times (1.98 \times 10^{-5})^{0.043}} = 1.22,$$

$$K_2 = \frac{0.833(0.0124 + 1/L')}{R'^{0.133}W'^{0.333}} = \frac{0.833 \times (0.0124 + 1/300)}{0.0118^{0.133} \times (1.98 \times 10^{5})^{0.333}}$$

$$= 0.871,$$

and

$$K_3 = \frac{1.12(1.52W'^{0.233} + 1/L')}{R'^{0.6}W'^{0.173}} = \frac{1.12 \times [1.52 \times (1.98 \times 10^{-5})^{0.233} + 1/300]}{0.0118^{0.6} \times (1.98 \times 10^{-5})^{0.173}}$$

$$= 13.1.$$

Then, using the rules given for the selection of X and K, as $R' < 0.15$

$$\text{as } X_2 > X_1 \quad X = X_1 = 7.11 \times 10^{-4},$$

and

$$\text{as } K_1 > K_2 \quad K = K_1 = 1.22.$$

Then critical heat flux ϕ_c can be calculated from eqn (D1)

$$\phi_c = XG(\lambda + K\Delta h_s)$$
$$= (7.11 \times 10^{-4}) \times 500 \times \{(1890 \times 10^3) + [1.22 \times (404.6 \times 10^3)]\}$$
$$= 8.47 \times 10^5 \text{ W/m}^2.$$

This is about 15 % higher than the value calculated by the procedure specifically for water. This difference is not unreasonably large and is of the order of the expected error in either calculation method. It may also be remembered that the tube length was outside the specified range for the water correlation. A check for the generalized correlation shows that the independent variables are within the specified range. The quality at the end of the tube for a heat flux of 8.47×10^5 W/m^2 is found to be 0.86.

Example 9 Film boiling

A tube of outside diameter 10 mm is made of material with emissivity 0.8. The outside temperature of the tube is 600 °C. The tube is submerged in saturated water at 1 bar. Calculate the heat flux when:

(a) there is no imposed liquid velocity and radiation is neglected;

(b) there is no imposed liquid velocity, and radiation is taken into account;

(c) there is an upward liquid velocity of 3 m/s, and radiation is neglected; and

(d) there is an upward liquid velocity of 3 m/s and radiation is taken into account.

The required physical properties are:

saturation temperature $T_{sat} = 100 °C = 373$ K;
latent heat of vaporization $\lambda = 2258 \times 10^3$ J/kg, and
liquid density $\rho_l = 958$ kg/m^3 (at 100 °C);

and the vapour properties at the mean film temperature (350 °C) are:

vapour density $\rho_g = 0.35$ kg/m^3,
vapour viscosity $\mu_g = 22 \times 10^{-6}$ N s/m^2,
vapour thermal conductivity $k_g = 0.045$ W/m K, and
vapour specific heat $C_{pg} = 2040$ J/kg K.

EXAMPLE 9 FILM BOILING   265

Solution

(a) *No liquid velocity, no radiation.* Equation (20.2) is used together with the latent heat correction equation given in the context of condensation (eqn (22.17)) with a constant of 0.68

$$\lambda' = \lambda + 0.68C_{pg}(T_{sat} - T_w)$$
$$= (2258 \times 10^3) + 0.68 \times 2040 \times (600 - 100)$$
$$= 2.952 \times 10^6 \text{ J/kg.}$$

In this equation the vapour specific heat is used in place of the liquid value. Then, using eqn (20.2) and putting $\Delta T_{sat} = T_{sat} - T_w = 600 - 100 = 500$ K

$$h = 0.62\left[\frac{\rho_g(\rho_l - \rho_g)g\lambda'k_g^3}{\mu_g \Delta T_{sat} d}\right]^{\frac{1}{4}}$$
$$= 0.62\left[\frac{0.35 \times (958 - 0.35) \times 9.81 \times (2.952 \times 10^6) \times 0.045^3}{22 \times 10^{-6} \times 500 \times 0.01}\right]^{\frac{1}{4}}$$
$$= 186 \text{ W/m}^2 \text{ K.}$$

The heat flux is then given by

$$\phi = h\Delta T_{sat} = 186 \times 500 = 9.3 \times 10^4 \text{ W/m}^2.$$

(b) *No liquid velocity, with radiation correction.* Equation (20.5) gives the radiative heat transfer coefficient. Note that here σ is Stefan's constant, and absolute temperatures must be used.

$$h_{radiative} = \frac{\varepsilon\sigma(T_w^4 - T_{sat}^4)}{T_w - T_{sat}} = \frac{0.8 \times (56.7 \times 10^{-9})(873^4 - 373^4)}{873 - 373}$$
$$= 51 \text{ W/m}^2 \text{ K.}$$

The total heat transfer coefficient is given by eqn (20.4) with the constant a equal to $\frac{3}{4}$

$$h = h_{\text{film boiling}} + \tfrac{3}{4}h_{radiative}$$
$$= 186 + (\tfrac{3}{4} \times 51) = 224 \text{ W/m}^2 \text{ K.}$$

The heat flux is then

$$\phi = h\Delta T_{sat} = 224 \times 500 = 11.2 \times 10^4 \text{ W/m}^2 \text{ K.}$$

The radiation correction has increased the heat flux by 20 %. At higher wall temperatures the correction is even larger.

(c) *Liquid velocity, no radiation.* Here the liquid velocity u is 3 m/s. The dimensionless group u^2/gd has a value of $(3 \times 3)/(9.81 \times 0.01) = 92$, and therefore eqn (20.6) is the relevant one. Again, using the corrected latent heat of vaporization λ'

$$h = 2.7\left(\frac{uk_g\rho_g\lambda'}{d\Delta T_{sat}}\right)^{\frac{1}{2}} = 2.7 \times \left[\frac{3 \times 0.045 \times 0.35 \times (2.952 \times 10^6)}{0.01 \times 500}\right]^{\frac{1}{2}}$$
$$= 451 \text{ W/m}^2\text{ K},$$

and the heat flux is

$$\phi = h\Delta T_{sat} = 451 \times 500 = 22.5 \times 10^4 \text{ W/m}^2.$$

Note that, in comparison with part (a), the effect of the liquid velocity has been to increase the heat transfer rate.

(d) *Liquid velocity, with radiation correction.* Now, using eqn (20.7),

$$h = h_{\text{film boiling}} + \tfrac{7}{8}h_{\text{radiative}}.$$

Then, taking $h_{\text{film boiling}} = 451$ from part (c), and $h_{\text{radiative}} = 51$ from part (b)

$$h = 451 + (\tfrac{7}{8} \times 51) = 496 \text{ W/m}^2\text{ K},$$

and the heat flux is

$$\phi = h\Delta T_{sat} = 496 \times 500 = 24.8 \times 10^4 \text{ W/m}^2.$$

The results for the heat fluxes are summarized in Table 25.5. The effect of the radiation correction increases very rapidly with increasing wall temperature.

TABLE 25.5 *Summary of heat flux results for Example 9.*

	No radiation	With radiation correction
No liquid velocity	9.3×10^4 W/m²	11.2×10^4 W/m²
Liquid velocity = 3 m/s	22.5×10^4 W/m²	24.8×10^4 W/m²

Example 10   Heat transfer to liquid drops

A drop of water of diameter $100 \, \mu$m and temperature $100\,°$C is carried upwards by a flow of steam with velocity 10 m/s and temperature $200\,°$C. The system pressure is 1 bar. Calculate the instantaneous value of dR/dt, where R is the drop radius:

The physical property data are:

liquid density	$\rho_l = 958 \text{ kg/m}^3$;
vapour density	$\rho_g = 0.46 \text{ kg/m}^3$;

EXAMPLE 10 HEAT TRANSFER TO LIQUID DROPS 267

vapour viscosity $\mu_g = 16 \times 10^{-6}$ N s/m²;
vapour thermal conductivity $k_g = 0.033$ W/m K;
vapour specific heat $C_{pg} = 1980$ J/kg K; and
latent heat of vaporization $\lambda = 2258 \times 10^3$ J/kg.

Solution

The basic equation is eqn (20.31). First, however, the gas Prandtl number is calculated from eqn (20.34)

$$Pr_g = \frac{\mu_g C_{pg}}{k_g} = \frac{(16 \times 10^{-6}) \times 1980}{0.033} = 0.96.$$

Equating the value of the drag force F given by eqn (20.36) and eqn (20.38), the velocity difference between the liquid drop and the gas is given by

$$u_g - u_d = \left[\frac{4}{3} \frac{D_d(\rho_1 - \rho_g)g}{\rho_g c_D} \right]^{\frac{1}{2}}.$$

This equation cannot be solved directly as the drag coefficient c_D is not known. Taking a first estimate of c_D as 0.44, which is the high Reynolds number value

$$u_g - u_d = \left[\frac{4}{3} \frac{(100 \times 10^{-6}) \times (958 - 0.46) \times 9.81}{0.46 \times 0.44} \right]^{\frac{1}{2}}$$

$$= 2.488 \text{ m/s}.$$

Then calculate the Reynolds number and the drag coefficient from eqns (20.33) and (20.37)

$$Re = \frac{\rho_g D_d(u_g - u_d)}{\mu_g} = \frac{0.46 \times (100 \times 10^{-6}) \times 2.488}{16 \times 10^{-6}} = 7.152,$$

and

$$c_D = \frac{24}{Re} + 0.44 = \frac{24}{7.152} + 0.44 = 3.796.$$

Then, repeating the calculations iteratively for $u_g - u_d$, Re, and c_D gives the values in Table 25.6. The relatively large number of iterations is caused by the very poor initial guess for c_D. The final value of $u_g - u_d$ will be about 0.32 m/s, and the final value of Re will be about 0.92.

Hence $u_d = 9.68$ m/s, as $u_g = 10$ m/s. The Nusselt number can now be calculated from eqn (20.31)

$$Nu = 2 + 0.6 Re^{\frac{1}{2}} Pr_g^{\frac{1}{3}} = 2 + (0.6 \times 0.92^{\frac{1}{2}} \times 0.96^{\frac{1}{3}})$$

$$= 2.57.$$

TABLE 25.6 *Results of the iteration in Example* 10

Value	2nd iteration	3rd iteration	4th iteration	5th iteration	6th iteration	7th iteration	8th iteration
c_D	3.796	10.296	16.673	21.097	23.676	25.056	25.763
$u_g - u_d$ (m/s)	0.847	0.514	0.404	0.359	0.339	0.330	0.325
Re	2.435	1.478	1.162	1.033	0.975	0.948	0.935
c_D	10.296	16.673	21.097	23.676	25.056	25.763	26.118

From eqn (20.32), the heat transfer coefficient is

$$h = \frac{Nu k_g}{D_d} = \frac{2.57 \times 0.033}{(100 \times 10^{-6})} = 847 \text{ W/m}^2 \text{ K.}$$

The heat transfer rate to the drop is given by eqn (20.35)

$$Q = \pi D_d^2 h (T_g - T_d) = \pi \times (100 \times 10^{-6})^2 \times 847 \times (200 - 100)$$
$$= 2.66 \times 10^{-3} \text{ W,}$$

and the mass of liquid vaporized per unit time is

$$\frac{Q}{\lambda} = \frac{2.66 \times 10^{-3}}{2258 \times 10^3} = 1.18 \times 10^{-9} \text{ kg/s.}$$

If the mass of the drop is m, this means that

$$\frac{dm}{dt} = -1.18 \times 10^{-9} \text{ kg/s,}$$

but

$$m = \tfrac{4}{3} \pi R^3 \rho_1,$$

and so

$$\frac{dm}{dt} = 4\pi R^2 \rho_1 \frac{dR}{dt},$$

or

$$\frac{dR}{dt} = \frac{dm}{dt} \frac{1}{4\pi R^2 \rho_1}$$
$$= \frac{(-1.18 \times 10^{-9})}{4\pi \times (50 \times 10^{-6})^2 \times 958} = -3.91 \times 10^{-5} \text{ m/s.}$$

Similarly, at other drop sizes the value of dR/dt can be calculated and thus the lifetime of the droplet can be found.

As the drop becomes small, then the Reynolds number also becomes small and the Nusselt number reaches a constant value of 2. For this case dR/dt is inversely proportional to R. Using this relationship the lifetime

EXAMPLE 11 CONDENSATION 269

of the droplet can be estimated as about 0.6 s. This seems very short, but it must be remembered that during this time the droplet will travel about 6 m.

Example 11 Condensation

(a) Benzene vapour at 1 bar condenses on a flat, vertical plate of height 0.3 m. The plate surface temperature is 60 °C. Calculate the average condensation heat transfer coefficient. The physical properties are:

saturation temperature at 1 bar $T_{sat} = 80$ °C;
liquid density $\rho_1 = 823$ kg/m³;
vapour density $\rho_g = 2.74$ kg/m³;
latent heat of vaporization $\lambda = 398 \times 10^3$ J/kg;
liquid specific heat $C_{pl} = 1880$ J/kg K;
liquid viscosity $\mu_1 = 321 \times 10^{-6}$ N s/m²; and
liquid thermal conductivity $k_1 = 0.131$ W/m K.

Solution

A corrected value of the latent heat is calculated from eqn (22.17) using a value of 0.68 for the constant in eqn (22.17), see page 215.

$$\lambda' = \lambda + 0.68C_{pl}(T_{sat} - T_w)$$
$$= 398 \times 10^3 + [0.68 \times 1880 \times (80 - 60)] = 424 \times 10^3 \text{ J/kg}.$$

The average heat transfer coefficient is then calculated from eqn (22.16) with λ' replacing λ

$$\Delta T_{sat} = T_{sat} - T_w = 80 - 60 = 20 \text{ K},$$

$$h = \frac{8^{\frac{1}{4}}}{3}\left[\frac{\rho_1(\rho_1 - \rho_g)g\lambda'k_1^3}{\mu_1\Delta T_{sat}L}\right]^{\frac{1}{4}}$$

$$= \frac{8^{\frac{1}{4}}}{3}\left[\frac{823 \times (823 - 2.74) \times 9.81 \times (424 \times 10^3) \times 0.131^3}{(321 \times 10^{-6}) \times 20 \times 0.3}\right]^{\frac{1}{4}}$$

$$= 1270 \text{ W/m}^2 \text{ K}.$$

(b) Benzene vapour at 1 bar condensed on the outside of a horizontal tube. The tube is very thin-walled, made of high thermal conductivity material, and of diameter 0.025 m. The inside of the tube is cooled by a flow of water at 25 °C. The heat transfer coefficient between the water and the inside of the wall is 1000 W/m² K. Calculate the condensation rate per metre of tube.

Solution

The temperature T_w of the tube wall is not known. By first guessing it at 60 °C, λ' can be calculated, as before, from eqn (22.17)

$$\lambda' = 424 \times 10^3 \text{ J/kg}.$$

Then the average condensation heat transfer coefficient h_c from eqn (22.18) is

$$h_c = 0.73 \left[\frac{\rho_1(\rho_1 - \rho_g)g\lambda'k_1^3}{\mu_1 \Delta T_{sat} d} \right]^{\frac{1}{4}}$$

$$= 0.73 \left[\frac{823 \times (823 - 2.74) \times 9.81 \times (424 \times 10^3) \times 0.131^3}{(321 \times 10^{-6}) \times (80 - T_w) \times 0.025} \right]^{\frac{1}{4}}$$

$$= \frac{3866}{(80 - T_w)^{\frac{1}{4}}} \text{ W/m}^2 \text{ K,}$$

where T_w is the wall temperature (°C). The heat transfer rate Q per metre is

$$Q = \pi d h_c (T_{sat} - T_w) = \pi \times 0.025 \times \frac{3866}{(80 - T_w)^{\frac{1}{4}}} \times (80 - T_w)$$

$$= 303.6 \times (80 - T_w)^{\frac{3}{4}} \text{ W/m}.$$

The heat transfer rate can also be calculated from the inside heat transfer coefficient h_i

$$Q = \pi d h_i (T_w - T_{water}) = \pi \times 0.025 \times 1000 \times (T_w - 25)$$

$$= 78.5 \times (T_w - 25).$$

These two equations for Q can be solved simultaneously (most easily by trial and error) to give values for Q and T_w. The solution is

$$T_w = 60.65 \text{ °C,}$$

and

$$Q = 2800 \text{ W/m}.$$

The rate of condensation $\dot{m}_c$ is then given by

$$\dot{m}_c = \frac{Q}{\lambda'} = \frac{2800}{424 \times 10^3} = 6.6 \times 10^{-3} \text{ kg/s m}.$$

Note that here λ' is again used, as it gives the heat extracted from the vapour per unit mass. Note also that in this case there is no need to recalculate λ' as it was calculated earlier on the basis of a wall temperature of 60 °C.

EXAMPLE 11 CONDENSATION 271

(c) A mixture of steam and air (9 parts by weight steam to 1 part by weight air) at 1 bar is condensing inside a horizontal tube. Find the condensing side heat transfer coefficient at the point where the quality is 0.15. The total flow rate is 0.01 kg/s, and the tube diameter is 0.02 m. A previous experiment with pure steam at the same total flow rate showed that when the quality was 0.15 the condensing heat transfer coefficient was 10 kW/m² K.

The gas phase, when the quality is 0.15, has the following physical properties:

density $\rho_g = 0.85$ kg/m³;
specific heat $C_{pg} = 1330$ J/kg K;
thermal conductivity $k_g = 0.028$ W/m K; and
viscosity $\mu_g = 18 \times 10^{-6}$ N s/m².

Solution

The basic equation is the Silver–Bell and Ghaly equation (eqn (22.34))

$$\frac{1}{h} = \frac{1}{h_f} + \frac{xC_{pg}\,\mathrm{d}T/\mathrm{d}h}{h_g},$$

where h is the actual heat transfer coefficient (W/m² K); h_f is the heat transfer coefficient across the condensate film (W/m² K); h_g is the heat transfer coefficient of the actual gas phase if no condensation occurred (W/m² K); and $\mathrm{d}T/\mathrm{d}h$ is the slope of the condensation curve (K kg/J).

The initial mixture composition and pressure are the same as the example given in § 22.9, hence the condensation curve is given in Fig. 22.15. From this figure, at a quality x of 0.15,

$$\frac{\mathrm{d}T}{\mathrm{d}h} = 8.2 \times 10^{-5} \text{ K kg/J.}$$

In order to calculate h_g, use the Dittus–Boelter equation

$$Nu_g = 0.023 Pr_g^{0.4} Re_g^{0.8},$$

where

$$Re_g = \frac{Gxd}{\mu_g},$$

$$Pr_g = \frac{\mu_g C_{pg}}{k_g},$$

and

$$Nu_g = \frac{h_g d}{k_g}.$$

Now

$$G = \frac{\text{mass flow rate}}{\text{cross-sectional area}} = \frac{0.01}{\pi(0.02)^2/4} = 31.83 \text{ kg/m}^2 \text{ s,}$$

so that

$$Re_g = \frac{31.83 \times 0.15 \times 0.02}{18 \times 10^{-6}} = 5305,$$

$$Pr_g = \frac{(18 \times 10^{-6}) \times 1330}{0.028} = 0.855,$$

$$Nu_g = 0.023 \times 0.855^{0.4} \times 5305^{0.8} = 20.6,$$

and

$$h_g = \frac{Nu_g k_g}{d} = \frac{20.6 \times 0.028}{0.02} = 28.9 \text{ W/m}^2 \text{ K.}$$

Now, substituting in eqn (22.34)

$$\frac{1}{h} = \frac{1}{10^4} + \frac{0.15 \times 1330 \times (8.2 \times 10^{-5})}{28.9}.$$

Therefore $h = 1500 \text{ W/m}^2 \text{ K.}$

Here the effect of the air mixed with the steam has been to reduce the condensation heat transfer coefficient from $10\,000 \text{ W/m}^2 \text{ K}$ to just $1500 \text{ W/m}^2 \text{ K}$. This dramatic reduction has occurred because, at a quality of 0.15, this particular mixture is mostly air and the condensation curve is relatively steep despite the apparently low value of dT/dh. All condensers must have particular attention paid to their venting arrangements so that non-condensable gas may be removed from the system. It is never safe to assume that non-condensable gas will not be present.

26

QUESTIONS

These questions are arranged in rough chapter order. References to equations and figures in the main text are given. It should be noted that the text is not sufficient to answer all the questions: for some, additional reading and thought are necessary. General references which are useful for the examples are Collier (1981), Hetsroni (1982), and Wallis (1969). These references, and others used in the questions, are given after the questions. Most of the necessary data on physical properties are given in the question or in the main text.

Chapter 1

(1) Air and water flow together in a tube of internal diameter 25 mm. The pressure is 1 bar and the phase densities are $\rho_g = 1.23 \text{ kg/m}^3$ and $\rho_1 = 1000 \text{ kg/m}^3$. The phase mass flow rates are 0.1 kg/s of air and 0.05 kg/s of water. The void fraction, $\alpha = 0.95$. Calculate:

(a) the total mass flux, G,
(b) the quality, x,
(c) the gas and liquid mass fluxes, G_g and G_1,
(d) the superficial gas and liquid velocities, V_g and V_1,
(e) the actual gas and liquid velocities, u_g and u_1.

Chapter 2

(2) From the Baker flow pattern map (see Fig. 2.6), draw the flow pattern boundaries on a graph of superficial velocities, V_g and V_1 for a flow of air and water in a horizontal tube of internal diameter 25 mm for conditions when the phase densities are as given in Question 1, the water viscosity, $\mu_1 = 10^{-3} \text{ N s/m}^2$, and the surface tension, $\sigma = 0.072 \text{ N/m}$. From your graph, and the results obtained in Question 1, find the predicted flow pattern for the flow in Question 1.

(3) For the flow of air and water in Question 1 use the Taitel and Dukler method for flow pattern determination (see § 2.4) to find the flow pattern for flow in a horizontal tube. For additional data, see Question 2. The air viscosity, $\mu_g = 17 \times 10^{-6} \text{ N s/m}^2$. The single phase pressure gradients can be calculated from the single phase friction factors, C_f

which can be assumed to be a function of the single phase Reynolds number, Re and given by:

$$C_f = 0.079 \, Re^{-0.25}. \tag{26.1}$$

Further information about the calculation of the single phase pressure gradients can be found in Chapter 25, Example 1.

(4) Repeat the calculation in Question 2 to find the predicted flow pattern for flow in a vertical tube. Use the flow pattern map of Hewitt and Roberts, see Fig. 2.7.

(5) The Taitel and Dukler method for flow pattern determination in a horizontal tube (see § 2.4) is based on the assumption that the flow is initially a smooth stratified flow, see Fig. 2.9. For an air and water flow in a tube of internal diameter 25 mm where the liquid and gas densities and viscosities are given in Questions 1, 2, and 3 and where the air superficial velocity is 10 m/s, calculate the water superficial velocity which will give $h_1/d = 0.5$, see Fig. 2.9. Assume that the air and water single phase friction factors, C_{fg} and C_{fl} are both constant and equal to 0.005.
Repeat the calculation with $h_1/d = 0.25$.

Chapter 3

(6) A series of equations for bubble rise velocity has been proposed by Peebles and Garber (1953) instead of the graphical method shown in Fig. 3.3. The equations have complicated ranges of applicability but the position can be summarized briefly by the statement that as the bubble size increases, the rise velocity is given successively by the five equations:

(a) Equation (3.3),

(b) $u_b = 0.136 \left[\dfrac{\rho_1}{\mu_1} \right]^{0.52} g^{0.76} D_e^{1.28},$ \hfill (26.2)

(c) $u_b = 1.91 \left[\dfrac{\sigma}{\rho_1 D_e} \right]^{0.5},$ \hfill (26.3)

(d) Equation (7.36),

(e) Equation (3.15).

Check that eqns (26.2) and (26.3) are dimensionally correct. Then

(a) Sketch the resulting form of the curve of u_b against D_c when plotted on a log–log graph.
(b) Prove that, at low pressure, all the above five equations can be

written conveniently in the form:

$$u_b = A \left[\frac{\sigma g}{\rho_1} \right]^a \left[\frac{\sigma}{\rho_1 D_e} \right]^b \left[\frac{D_e^2 g \rho_1}{\mu_1} \right]^c \tag{26.4}$$

and find A, a, b, and c for each equation.

(c) Prove that the equivalent diameters at which one equation gives way to the next (assuming $\rho_1 \gg \rho_g$) can be written in the form:

$$D_e = B \left[\frac{\sigma}{\rho_1 g} \right]^d \left[\frac{\mu_1^2}{g \rho_1^2} \right]^e \tag{26.5}$$

(d) Find the detailed shape of the curve of u_b as a function of D_e for benzene at 1 bar. Compare with the results from Fig. 3.3 which are given in Fig. 25.1. The properties of saturated benzene at 1 bar are given in Example 2 in Chapter 25.

(7) If the rising velocity of a plug, u_p depends on the six variables:

tube diameter, d
acceleration due to gravity, g
liquid density, ρ_1
gas density, ρ_g
liquid viscosity, μ_1
surface tension, σ

explain why Fig. 3.8 expresses the rising velocity in terms of a relationship between three dimensionless groups rather than four.

(a) Write expressions for the ratio of forces due to liquid inertia, liquid viscosity, and surface tension to the buoyancy force. Relate these ratios to those used in Fig. 3.8.

(b) Using physical arguments why will the plug not move if the Eotvos number, Eo, is of order unity? (In fact the critical Eotvos number for movement is 3.37.)

(c) When liquid inertia is the dominant force ($Eo^3/M \gg 1$) the Froude number is equal to 0.35, and when liquid viscosity is the dominant force ($Eo^3/M \ll 1$) the ratio $\dfrac{u_p \mu_1}{gd^2(\rho_1 - \rho_g)}$ is equal to 0.01. Using these statements and the critical Eotvos number given in part (b), can you suggest a general form for the dependence of the Froude number on two other dimensionless groups! Here M is the Morton number.

(8) Air and water at 1 bar are contained in a tube of diameter 50 mm.

(a) Calculate the diameter of a plug in a tube emptying experiment, see Fig. 3.12.

(b) Air is blown into the bottom of the tube at a volume flow rate of $2 \times 10^{-4}\,\mathrm{m^3/s}$ as shown in Fig. 3.13. Calculate the void fraction and the actual velocity of the rising plugs. Take account of the velocity profile across the tube diameter.

Chapter 4

(9) Prove eqn (4.6). If the film thickness, m is used instead of r_i and the distance from the wall, y is used instead of r, write an alternative form of eqn (4.6) and simplify it for the case where $m/r_0 \ll 1$, and $y/r_0 \ll 1$ to arrive at eqn (4.7).

(10) Section 4.4 gives an example of the use of the triangular relationship. If in this example, the gas density is $1.23\,\mathrm{kg/m^3}$, the gas viscosity is $17 \times 10^{-6}\,\mathrm{N\,s/m^2}$, and gas velocity is $30\,\mathrm{m/s}$, calculate the interfacial shear stress from the experimental value of the film thickness using the interfacial roughness correlation, see § 4.5 and eqn (4.28). Compare your result with the actual shear stress calculated from the pressure gradient. How would you calculate values of both the film thickness and the interfacial shear stress which would satisfy the triangular relationship and the interfacial roughness correlation?

Chapter 5

(11) With the data given in Question 1, calculate the homogeneous void fraction for the flow. Comment on the result.

(12) For the flow in Question 1, calculate the values of $(-\mathrm{d}p/\mathrm{d}z)_\mathrm{g}$, $(-\mathrm{d}p/\mathrm{d}z)_\mathrm{go}$, $(-\mathrm{d}p/\mathrm{d}z)_\mathrm{l}$, $(-\mathrm{d}p/\mathrm{d}z)_\mathrm{lo}$, and, assuming the homogeneous viscosity if given by eqn (5.13), $(-\mathrm{d}p/\mathrm{d}z)_\mathrm{F}$. Hence calculate the two phase multipliers ϕ_g^2, ϕ_go^2, ϕ_l^2, and ϕ_lo^2. Viscosity data for the gas and the liquid are given in Questions 2 and 3.

(13) Prove eqns (5.30), (5.31), and (5.32).

Chapter 6

(14) Derive eqn (6.15) from eqn (6.12). What are the reasons why the frictional pressure gradients calculated from eqns (6.8) and (6.15) are different? Which of the frictional terms will be larger in a vertical upflow where the accelerational terms are small?

(15) For the flow in Question 1, calculate the void fraction using:

(a) Zivi's correlation, and
(b) Chisholm's correlation.

(16) For the flow in question 1, calculate the momentum flux for:

(a) homogeneous flow,
(b) separated flow, using the void fraction from Zivi's correlation, and
(c) separated flow, using the void fraction from Chisholm's correlation.

Account for the fact that although the three cases above all have void fractions almost equal to unity, the momentum flux in the separated flow cases is significantly different from that in the homogeneous flow case.

(17) From the results of question 12, calculate the value of C for the homogeneous model from eqn (6.39). Also calculate C from eqn (6.47) and account for any discrepancy.

Selecting the appropriate value of C from Table 6.2, calculate the frictional pressure gradient from the Lockhart–Martinelli correlation.

Chapter 7

(18) In a bubbly flow the drift flux j_{gl} is given by eqn (7.17). There is no liquid flow, and the gas superficial velocity is gradually increased from a low value. Show that there are, initially, theoretically two values of the void fraction possible, and that these two solutions approach each other as the gas velocity increases. Find the gas velocity (in terms of u_b) and the void fraction at which the two solutions coincide. Why would this solution be difficult to obtain experimentally?

(19) Prove eqns (7.18) and (7.19). For a bubbly flow with $u_b = 0.2\,\text{m/s}$ and $V_g = 0.02\,\text{m/s}$, find the liquid velocity necessary to cause flooding, and the void fraction at flooding.

(20) Calculate the distribution parameter, C_0 from eqn (7.33) for flow in a round tube if both the void fraction and the velocity profiles are parabolic.

Chapter 8

(21) In eqn (8.12) it is assumed that the liquid is incompressible. If $(\partial v_l/\partial p)_s$ is assumed to be equal to $(\partial v_l/\partial p)_T$, and using the adiabatic

compressibility κ where:

$$\kappa = -\frac{1}{v_1}\left[\frac{\partial v_1}{\partial p}\right]_T \qquad (26.6)$$

derive a version of eqn (8.15) which takes account of liquid compressibility effects. One of the solutions of the example in § 8.5 gives the quality at the choking plane as 2.8 % and the pressure at the choking plane as 30 bar. Assess the relative importance of the liquid compressibility term for this example with these values of quality and pressure. Assume that the relevant value of κ is $4 \times 10^{-10}\,\mathrm{m^2/N}$.

(22) Figure 8.5 gives the results of a trial and error solution for critical flow of steam–water using the homogeneous equilibrium model. Verify the results by calculating the mass flux when the pressure at the choking plane is 28 bar, 30 bar, and 32 bar.

(23) Prove that a slip ratio of $(\rho_1/\rho_g)^{1/2}$ minimizes the momentum flux of a two-phase flow.

Chapter 9

(24) From the data in § 9.2 estimate the mean separation of the waves in annular flow, and the length of the waves at half the maximum amplitude (see Fig. 9.6). What is the important difference between these waves and conventional deep water waves?.

(25) Use the data in Fig. 9.14 to test the correlation of Ishii and Mishima for entrained liquid fraction in annular flow.

(26) Using the data given in § 9.6 for typical air–water values in an annular flow, calculate the droplet deposition mass transfer coefficient, k for a droplet diameter, D_d of 100 μm.
Calculate also the deposition length required to reduce the entrained liquid flow rate to 10 % of its original value if $d = 0.025$ m, $\rho_g = 2\,\mathrm{kg/m^3}$, and $G_g = 100\,\mathrm{kg/m^2 s}$.

Chapter 11

(27) Section 11.7 gives correlations for flooding due to Wallis and Pushkina and Sorokin. For air–water flow with $\rho_1 = 1000\,\mathrm{kg/m^3}$ and $\rho_g = 2\,\mathrm{kg/m^3}$, what tube diameter gives the same flooding gas velocity if the liquid velocity is low? Take $C = 0.8$. If the tube diameter is lower

than this value, which correlation gives the most conservative (lower) result for the flooding gas velocity?

(28) A bottle has volume V and neck internal diameter d. It is full of water, and emptied by turning it upside down. If the emptying rate is controlled by flooding, and $C = 0.8$ in the Wallis correlation, prove that the time t for the bottle to empty is given by:

$$t = \frac{4V}{\pi d^2 C^2} \frac{[\rho_l^{\frac{1}{4}} + \rho_g^{\frac{1}{4}}]^2}{[gd(\rho_l - \rho_g)]^{\frac{1}{2}}}. \tag{26.7}$$

Hence find the emptying time for a bottle with $V = 1$ litre and $d = 20$ mm.

(29) Repeat the previous question if the bottle emptying is now controlled by the entry of plug flow bubbles into the bottle. If these bubbles are inertia controlled, show that:

$$t = \frac{4V}{0.35 \pi d^2 (gd)^{\frac{1}{2}}} \left[\frac{\rho_l}{\rho_l - \rho_g} \right]^{\frac{1}{2}}. \tag{26.8}$$

Again find the emptying time for $V = 1$ litre and $d = 20$ mm.

Chapter 12

(30) A liquid enters a vertical tube and flows upwards. The tube is heated by a source of constant heat flux. Find a version of eqn (5.34) which is applicable for the situation where the liquid enters the vertical tube subcooled (i.e. below its boiling point). Express the equation in terms of the total length of the tube, L and the length over which the liquid is subcooled, L_s. Then:

(a) by a heat balance prove that the quality at the end of the tube, x_o is given by:

$$x_o = \frac{4L\phi}{dG\lambda} - \frac{\Delta h_s}{\lambda} \tag{26.9}$$

where ϕ = heat flux (assumed constant along the tube), λ = latent heat of vaporization, and Δh_s = enthalpy of subcooling at entry to the tube (this is the saturation liquid enthalpy − actual liquid enthalpy).

(b) Prove that the subcooling length L_s is given by:

$$L_s = \frac{dG \Delta h_s}{4\phi}. \tag{26.10}$$

(c) For the case where $\rho_l = 823$ kg/m³, $\rho_g = 2.74$ kg/m³, $\lambda = 398$ kJ/kg,

$\Delta h_s = 100 \text{ kJ/kg}$, $L = 10 \text{ m}$, $d = 0.01 \text{ m}$, $\phi = 30 \text{ kW/m}^2$, and taking $C_{\text{flo}} = 0.005$, find the three values of the mass flux (as in Fig. 12.2) which give a pressure drop of $1.25 \times 10^5 \text{ N/m}^2$. Which of these are potentially stable operating points? Disregard the effect of pressure on any of the physical properties.

Chapter 13

(31) Under what conditions will phase inversion (see § 13.2) occur in a bend if the void fraction is given by:

(a) Chisholm's equation (see § 6.5), or
(b) the minimum momentum flux condition, see question 23?

Chapter 14

(32) A spherical cavity in a liquid is at a pressure Δp above the liquid pressure. The cavity expands from a radius R_o to R. Show that the mechanical work done, W is given by:

$$W = \frac{4\pi}{3}(R^3 - R_o^3)\Delta p \qquad (26.11)$$

and also by:

$$W = \frac{\rho_1}{2}\int_R^\infty 4\pi r^2 \dot{r}^2 \, dr. \qquad (26.12)$$

Relate $\dot{r}$ to $\dot{R}$ and substitute in eqn (26.12). Perform the integration in eqn (26.12) and equate the result to eqn (26.11). Then differentiate to eliminate R_o and hence prove the Rayleigh equation of bubble growth:

$$R\ddot{R} + \tfrac{3}{2}\dot{R}^2 = \frac{\Delta p}{\rho_1}. \qquad (26.13)$$

Find the only solution of the form $R = at^n$ and find t and a. Why is this solution only valid at short times? What controls the rate of bubble growth at large times?

(33) Calculate the heater temperature necessary to start pool boiling in water at 20 bar if the cavity size available is of diameter $8 \, \mu\text{m}$. The physical properties of saturated water at 20 bar can be found in Example 8 in Chapter 25.

(34) Calculate the relationship between heat transfer coefficient h

(W/m² K) and heat flux ϕ (W/m²) for water at 20 bar in pool boiling. Express the relationship in the form $h = a\phi^b$. The physical properties can be found in Example 8 in Chapter 25; the critical pressure of water is 221.1 bar. Use the following methods:

(a) Mostinski method—eqns (14.23) and (14.24).
(b) Cooper method—eqn (14.25), take $A = 40$, and the roughness, $\varepsilon = 1\,\mu m$.
(c) Rohsenow method—eqn (14.22), take $n = 0.33$, $m = 0.7$, and $C_{sf} = 0.025$.
(d) Foster–Zuber method—eqn (16.13). Here use the Clausius–Clapeyron equation to eliminate Δp_{sat} from the equation.

The result in part (d) is significantly different from the other results. At what heat flux does the Foster–Zuber method give the same value of the heat transfer coefficient as the Mostinski method?

Thom et al. (1965) produced a correlation for heat transfer to water at high pressures in areas where the heat transfer was dominated by nucleate boiling effects. The correlation, which is really only valid between 52 bar and 138 bar, is (p is measured in bar in this equation):

$$\Delta T_{sat} = \frac{0.0225\phi^{0.5}}{e^{p/86.9}}. \tag{26.14}$$

Compare the results of this equation with those of the Foster–Zuber method.

Chapter 15

(35) Explain why a bottle full of liquid which has the neck covered by fine gauze will not empty when turned carefully upside down. For a bottle filled with water emptying into air, estimate the order of magnitude of the gauze mesh which will just allow emptying.

(36) Explain why a flag 'flutters' in the wind, and does not adopt a flat, stationary position.

(37) There are few data points available for critical heat flux in pool boiling on bundles of horizontal tubes. Explain why data obtained from experiments such as that illustrated in Fig. 14.1 are of little use.

(38) A vertical tube is uniformly heated with a heat flux ϕ, and is closed at its lower end. The tube is fed with saturated water at 20 bar from a pool above the top of the tube. Calculate the maximum possible heat flux, if the tube length and internal diameter are 0.5 m and 0.01 m

respectively. The properties of saturated water and steam at 20 bar can be found in Chapter 25, Example 8. Compare your result with the values tabulated in Table 15.1.

Chapter 16

(39) Saturated water at 20 bar flows in a heated tube. The heat flux is 5×10^5 W/m^2. Calculate the minimum wall temperature necessary for the onset of nucleation and the cavity size which will first become active. The properties of water at 20 bar are given in Chapter 25, Example 8. Assume that a full range of cavity sizes are present on the tube wall. If the wall temperature is now 488 K, find the range of cavity sizes which are active.

(40) Water at 470 K and 20 bar flows along a tube of internal diameter 0.01 m at a mass flux of 1000 kg/m^2 s. If the heat flux is 5×10^5 W/m^2, calculate the wall temperature. Take the properties of water to be the saturated properties from Chapter 25, Example 8. Calculate the nucleate boiling heat transfer coefficient h_{NB} (W/m^2 K) from:

$$h_{NB} = 60\phi_{NB}^{0.5} \tag{26.15}$$

where ϕ_{NB} is the nucleate boiling part of the heat flux (W/m^2).

Chapter 17

(41) Using the simplified version of the MacBeth correlation for critical heat flux in water:

$$\phi_c = A\lambda G^{\frac{1}{2}}(1-x) \tag{26.16}$$

where $A = 0.25$ (kg/m^2 s)$^{\frac{1}{2}}$, calculate the critical heat flux and the critical quality for the situation in Example 8 in Chapter 25.

(42) Using the simplified version of the MacBeth correlation for critical heat flux, see eqn (26.16), verify *all* the parametric trends shown in Fig. 17.5.

(43) Saturated water enters a vertical tube of length L and internal diameter d which is non-uniformly heated so that the distribution of the heat flux ϕ is given by:

$$\phi = \phi_{max}\left[1 - \frac{z}{L}\right] \tag{26.17}$$

where z is the distance from the bottom of the tube. Show that, using the critical heat flux correlation eqn (26.16), the critical heat flux first occurs at a distance z_c from the bottom of the tube, where z_c is given by:

$$z_c = L - \frac{dG^{\frac{1}{2}}}{4A} \tag{26.18}$$

Explain what will happen if $G > \dfrac{16A^2L^2}{d^2}$.

(44) In the previous question, if the critical heat flux occurs part way up the tube, show that the value of ϕ_{max} corresponding to the critical heat flux is given by

$$\phi_{max} = \frac{A\lambda G^{\frac{1}{2}}}{\dfrac{dG^{\frac{1}{2}}}{8AL} + \dfrac{2AL}{dG^{\frac{1}{2}}}}. \tag{26.19}$$

For the particular case where $dG^{\frac{1}{2}} = AL$, find the heat flux averaged over the whole tube at the critical condition. Compare this with the critical heat flux in a uniformly heated tube.

(45) The Bowring critical heat flux correlation is given in eqn (17.5). What are the necessary conditions on A, B, and C in eqn (17.5) for the Bowring correlation to become identical to the approximate MacBeth correlation given by eqn (26.16)? Examine Appendix C and consider whether such conditions are likely ever to be approximately true.

Chapter 19

(46) Derive the overall (gas and liquid) energy equation analogue of eqn (19.48) in § 19.10.

Chapter 20

(47) Equation (20.2) and eqn (20.6) are two equations for film boiling heat transfer coefficients. Inequality (20.9) is a way of choosing which equation to use. In what way is this consistent with choosing the one which will give a higher value of the heat transfer coefficient?

Repeat the calculation in § 20.2 of the film boiling heat transfer coefficient, but with an upwards liquid velocity of 1.5 m/s imposed upon the situation.

(48) A horizontal cylinder of diameter 0.02 m is heated in water at 1 bar until the critical heat flux of 1.25 MW/m² is exceeded. Estimate the

TABLE 26.1   *Physical properties of steam at 1 bar*

T (°C)	100	200	300	400	500	600	700	800	900	1000
k_g (mW/m K)	25	33	43	55	67	80	93	107		
μ_g (μNs/m^2)	12	16	20	24	29	33	37	40		
ρ_g (kg/m^3)	0.59	0.46	0.38	0.32	0.28	0.25	0.22	0.20	0.18	0.17

temperature of the cylinder taking radiative heat transfer into account as well as film boiling. Take data on physical properties from § 20.2 and from Table 26.1. Assume an emissivity of 1.0.

Chapter 22

(49) Prove eqn (22.17).

(50) Water at 20 °C flows through a horizontal pipe of external diameter 0.02 m and internal diameter 0.015 m. Steam at 1 bar condenses on the outside of the tube. The tube metal thermal conductivity is 50 W/m K. Calculate the heat transfer rate per unit length of pipe if the heat transfer coefficient between the water and the inside of the tube wall is 1250 W/m^2 K. Calculate also the tube inside and outside wall temperatures.

What will be the most effective way of increasing the heat transfer rate substantially? What heat transfer rate could be reached without difficulty? You may assume that the inside heat transfer coefficient is given by the Dittus–Boelter equation:

$$Nu = 0.023 \, Re^{0.8} \, Pr^{0.4} \qquad (26.20)$$

References

COLLIER, J. G. (1981). *Convective boiling and condensation.* McGraw-Hill, New York.

HETSRONI, G. (ed.) (1982). *Handbook of multiphase systems.* McGraw-Hill, New York.

PEEBLES, F. N. and GARBER, H. J. (1953). Studies on the motion of gas bubbles in liquids. *Chem. Eng. Prog.,* **49,** 88–97.

THOM, J. R. S., WALKER, W. M., FALLON, T. A., and REISING, G. F. S. (1965). Boiling in sub-cooled water during flow up heated tubes or annuli. *Proc. Instn. Mech. Engnrs.,* **180,** (Part 3c), 226–246.

WALLIS, G. B. (1969). *One dimensional two-phase flow.* McGraw-Hill, New York.

APPENDIX A
CISE Correlation for void fraction

The correlation of Premoli *et al.* (1970), usually known as the CISE correlation is a correlation in terms of the slip ratio S. The void fraction/α is then given by

$$\alpha = \frac{1}{1 + \left(S \dfrac{1-x}{x} \dfrac{\rho_g}{\rho_l} \right)} \tag{A1}$$

The slip ratio is then given by

$$S = 1 + E_1 \left(\frac{y}{1 + yE_2} - yE_2 \right)^{0.5} \tag{A2}$$

where

$$y = \frac{\beta}{1 - \beta}, \tag{A3}$$

$$\beta = \frac{\rho_l x}{\rho_l x + \rho_g(1 - x)}, \tag{A4}$$

$$E_1 = 1.578 \, Re^{-0.19} \left(\frac{\rho_l}{\rho_g} \right)^{0.22}, \tag{A5}$$

$$E_2 = 0.0273 \, We \, Re^{-0.51} \left(\frac{\rho_l}{\rho_g} \right)^{-0.08}, \tag{A6}$$

$$Re = \frac{Gd}{\mu_l}, \tag{A7}$$

and

$$We = \frac{G^2 d}{\sigma \rho_l}. \tag{A8}$$

Here: x is the quality; ρ_g is the gas density (kg/m^3); ρ_l is the liquid density (kg/m^3); d is the tube diameter (m); μ_l is the liquid viscosity (N s/m^2); G is the total (liquid + gas) mass flux (kg/m^2 s); and σ is the surface tension (N/m).

Note that the definition of the Weber number We is not the same as that used in Appendix B.

Reference

PREMOLI, A., FRANCESCO, D., and PRINA, A. (1970). An empirical correlation for evaluating two-phase mixture density under adiabatic conditions. *European Two-Phase Flow Group Meeting, Milan.*

285

APPENDIX B

Friedel correlation for frictional two-phase pressure gradient

The Friedel (1979) correlation is probably the most accurate generally available correlation for the frictional two-phase pressure gradient. It is written in terms of a two-phase multiplier

$$\phi_{lo}^2 = \frac{(-dp/dz)_F}{(-dp/dz)_{lo}}, \tag{B1}$$

where $(-dp/dz)_F$ is the frictional pressure gradient (N/m^3) in the two-phase flow, and $(-dp/dz)_{lo}$ is the frictional pressure gradient (N/m^3) in single-phase liquid flow with the same mass flow rate as the total two-phase flow rate. Then

$$\phi_{lo}^2 = E + \frac{3.24\, FH}{Fr^{0.045} We^{0.035}}, \tag{B2}$$

$$E = (1-x)^2 + x^2 \frac{\rho_1 C_{fgo}}{\rho_g C_{flo}}, \tag{B3}$$

$$F = x^{0.78}(1-x)^{0.224}, \tag{B4}$$

$$H = \left(\frac{\rho_1}{\rho_g}\right)^{0.91} \left(\frac{\mu_g}{\mu_1}\right)^{0.19} \left(1 - \frac{\mu_g}{\mu_1}\right)^{0.7} \tag{B5}$$

$$Fr = \frac{G^2}{g\, d\rho_h^2}, \tag{B6}$$

$$We = \frac{G^2 d}{\sigma \rho_h}, \tag{B7}$$

and

$$\rho_h = \left(\frac{x}{\rho_g} + \frac{1-x}{\rho_1}\right)^{-1}. \tag{B8}$$

Here x is the quality; ρ_g is the gas density (kg/m^3); ρ_1 is the liquid density (kg/m^3); C_{fgo} and C_{flo} are the friction factors for the total mass flux flowing with the gas and liquid properties respectively; μ_g is the gas viscosity $(N\,s/m^2)$; μ_1 is the liquid viscosity $(N\,s/m^2)$; σ is the surface tension (N/m); and g is the acceleration due to gravity $(=9.81\ m/s^2)$.

Note that the definition of the Weber number We is not the same as that used in Appendix A.

The correlation is applicable to vertical upflow and to horizontal flow. A very

286

similar correlation is available for vertical downflow. The method is known not to work very well when the viscosity ratio (μ_l/μ_g) exceeds 1000.

Reference

FREIDEL, L. (1979). Improved friction pressure drop correlations for horizontal and vertical two-phase flow. *European Two-phase Flow Group Meeting, Ispra, Italy*.

APPENDIX C

Bowring correlation for critical heat flux of water in a vertical tube

The correlation of Bowring (1972) for critical heat flux is applicable for the following conditions:

fluid—steam–water systems only;
flow direction—vertical upflow;
pressure $p = 2$–190 bar;
tube diameter $d = 2$–45 mm;
tube length $L = 0.15$–3.7 m; and
total mass flux $G = 136$–18 600 kg/m^2 s.

Inevitably, not all the possible combinations of variables inside these ranges have been covered by experimental data.

The basic equation for the critical heat flux ϕ_c (W/m^2) is

$$\phi_c = \frac{A' + 0.25\, dG\, \Delta h_s}{C' + L}, \tag{C1}$$

where Δh_s is the inlet subcooling (J/kg).
A' and C' are given by

$$A' = \frac{0.5792\lambda dG F_1}{1 + 0.0143 F_2 d^{0.5} G} \tag{C2}$$

and

$$C' = \frac{0.077 F_3 dG}{1 + 0.347 F_4 (G/1356)^n}, \tag{C3}$$

where λ is the latent heat of vaporization (J/kg) and F_1, F_2, F_3, F_4, and n are functions of a non-dimensional pressure p'.

$$p' = \frac{p}{69}, \tag{C4}$$

where p is the system pressure in bar (these units must be used).

$$n = 2.0 - 0.5p'. \tag{C5}$$

For $p' < 1$

$$F_1 = \frac{\{p'^{18.942} \exp[20.8(1 - p')]\} + 0.917}{1.917}, \tag{C6}$$

$$\frac{F_1}{F_2} = \frac{\{p'^{1.316} \exp[2.444(1 - p')]\} + 0.309}{1.309}, \tag{C7}$$

$$F_3 = \frac{\{p'^{17.023} \exp[16.658(1 - p')]\} + 0.667}{1.667}, \tag{C8}$$

$$\frac{F_4}{F_3} = p'^{1.649}. \tag{C9}$$

For $p' > 1$

$$F_1 = p'^{-0.368} \exp[0.648(1 - p')], \tag{C10}$$

$$F_2 = p'^{-0.448} \exp[0.245(1 - p')], \tag{C11}$$

$$F_3 = p'^{0.219}, \tag{C12}$$

$$\frac{F_4}{F_3} = p'^{1.649}. \tag{C13}$$

The root-mean-square error found by Bowring was about 7 %, however because the correlation is wholly empirical, conditions not representative of those in the original set of data used to derive the correlation can give rise to much larger errors.

Reference

BOWRING, R. W. (1972). A simple but accurate round tube, uniform heat flux, dryout correlation for pressure in the range 0.7–17 MN/m^2. *AEEW-R789*.

APPENDIX D

Katto generalized correlation for critical heat flux in vertical tubes

Katto and Ohne (1984) have correlated the critical heat flux ϕ_c (W/m^2) by the basic equation

$$\phi_c = XG(\lambda + K \, \Delta h_s), \tag{D1}$$

where G is the total mass flux (kg/m^2 s); λ is the latent heat of vaporization (J/kg); and Δh_s is the inlet subcooling (J/kg).

X and K are functions of three dimensionless groups

$$L' = \frac{L}{d}, \tag{D2}$$

$$R' = \frac{\rho_g}{\rho_l}, \tag{D3}$$

$$W' = \frac{\sigma \rho_l}{G^2 L}, \tag{D4}$$

where L is the tube length (m); d is the tube diameter (m); ρ_g is the gas density (kg/m^3); ρ_l is the liquid density (kg/m^3); and σ is the surface tension (N/m).

Five values of X can then be calculated

$$X_1 = \frac{CW'^{0.043}}{L'}, \tag{D5}$$

$$X_2 = \frac{0.1R'^{0.133}W'^{0.333}}{1 + 0.0031L'}, \tag{D6}$$

$$X_3 = \frac{0.098R'^{0.133}W'^{0.433}L'^{0.27}}{1 + 0.0031L'}, \tag{D7}$$

$$X_4 = \frac{0.0384R'^{0.6}W'^{0.173}}{1 + 0.28W'^{0.233}L'}, \tag{D8}$$

and

$$X_5 = \frac{0.234R'^{0.513}W'^{0.433}L'^{0.27}}{1 + 0.0031L'}, \tag{D9}$$

where the value of C in eqn (D5) is given by

for $L' < 50$ $C = 0.25$ (D10)

for $50 < L' < 150$ $C = 0.25 + 0.0009(L' - 50)$ (D11)

for $L' > 150$ $C = 0.34$.   (D12)

Three values of K can be calculated

$$K_1 = \frac{0.261}{CW'^{0.043}},$$ (D13)

$$K_2 = \frac{0.833(0.0124 + 1/L')}{R'^{0.133}W'^{0.333}}$$ (D14)

$$K_3 = \frac{1.12(1.52W'^{0.233} + 1/L')}{R'^{0.6}W'^{0.173}}.$$ (D15)

Then the applicable values of X and K are chosen according to the following rules.

For $R' < 0.15$:

if $X_1 < X_2$	then	$X = X_1$;
if $X_1 > X_2$ and $X_2 < X_3$	then	$X = X_2$;
if $X_1 > X_2$ and $X_2 > X_3$	then	$X = X_3$;
if $K_1 > K_2$	then	$K = K_1$;
if $K_1 < K_2$	then	$K = K_2$.

For $R' > 0.15$:

if $X_1 < X_5$	then	$X = X_1$;
if $X_1 > X_5$ and $X_5 > X_4$	then	$X = X_5$;
if $X_1 > X_5$ and $X_5 < X_4$	then	$X = X_4$;
if $K_1 > K_2$	then	$K = K_1$;
if $K_1 < K_2$ and $K_2 < K_3$	then	$K = K_2$;
if $K_1 < K_2$ and $K_2 > K_3$	then	$K = K_3$.

The range of data used by Katto was:

$L = 0.01$ m to 8.8 m;
$d = 0.001$ m to 0.038 m;
$L' = 5$ to 880;
$R' = 0.0003$ to 0.41; and
$W' = 3 \times 10^{-9}$ to 2×10^{-2}.

Typical errors in the calculated critical heat flux are around 20 %; it is thus less accurate than the Bowring correlation for water (see Appendix C) but has been tested for a range of experimental fluids.

References

KATTO, Y. and OHNE, H. (1984). An improved version of the generalized correlation of critical heat flux for convective boiling in uniformly heated vertical tubes'. *Int. J. Heat Mass Transfer* **27**, 1641–8.

INDEX